Photonics Meeting 2021 (PM21)

Journal of Physics: Conference Series Volume 2075

Online
15 – 16 September 2021

ISBN: 978-1-7138-5046-5
ISSN: 1742-6588

Printed from e-media with permission by:

Curran Associates, Inc.
57 Morehouse Lane
Red Hook, NY 12571

CURRAN ASSOCIATES INC.
proceedings
.com

Some format issues inherent in the e-media version may also appear in this print version.

No changes have been made to the content of these proceedings. There may be changes to pagination and minor adjustments for aesthetics.

Printed with permission by Curran Associates, Inc. (2023)

For permission requests, please contact the Institute of Physics at the address below.

Institute of Physics
Dirac House, Temple Back
Bristol BS1 6BE UK

Phone: 44 1 17 929 7481
Fax: 44 1 17 920 0979

techtracking@iop.org

Additional copies of this publication are available from:

Curran Associates, Inc.
57 Morehouse Lane
Red Hook, NY 12571 USA
Phone: 845-758-0400
Fax: 845-758-2633
Email: curran@proceedings.com
Web: www.proceedings.com

TABLE OF CONTENTS

Preface

The 4th Photonics Meeting 2021 (PM21) is an annual event organized by the Optical Society of Malaysia (OSM) in celebrating UNESCO International Day of Light. In conjunction with Photonics Meeting 2021, The 2nd Webinar on Engineering Education 4.0: Lasers, Optics and Photonics Technology was organized by Universiti Teknologi MARA Cawangan Terengganu in a collaboration with Universitas Airlangga, Surabaya. This program is a high impact program which allow researchers and scientists to share advances in optical engineering and applications, nanotechnology, while exchanging scientific knowledge among participants. The program has gather experts in Lasers, Optics and Photonics field to contribute in delivering ideas and experiences to all participants involved. This was held on 15th and 16th September 2021 virtually due to Movement Control Order (MCO) by the government. It was host by Universiti Teknologi MARA, Dungun, Terengganu, Malaysia. The speakers have presented their work using Cisco Webex of maximum 10 minutes and 5 minutes of question and answer. This two-day event consists of a conference and a webinar that receives nearly 21 participants from the photonics and optics field. It has been successfully conducted without any difficulties. A compilation of articles that represent the significant results of the conference has been selected to publish in this volume.

Editors
MOHD HAFIZ BIN JALI
ANAS ABDUL LATIFF
SULAIMAN WADI HARUN

Journal of Physics: Conference Series **2075** (2021) 011001 doi:10.1088/1742-6596/2075/1/011001

Day 1

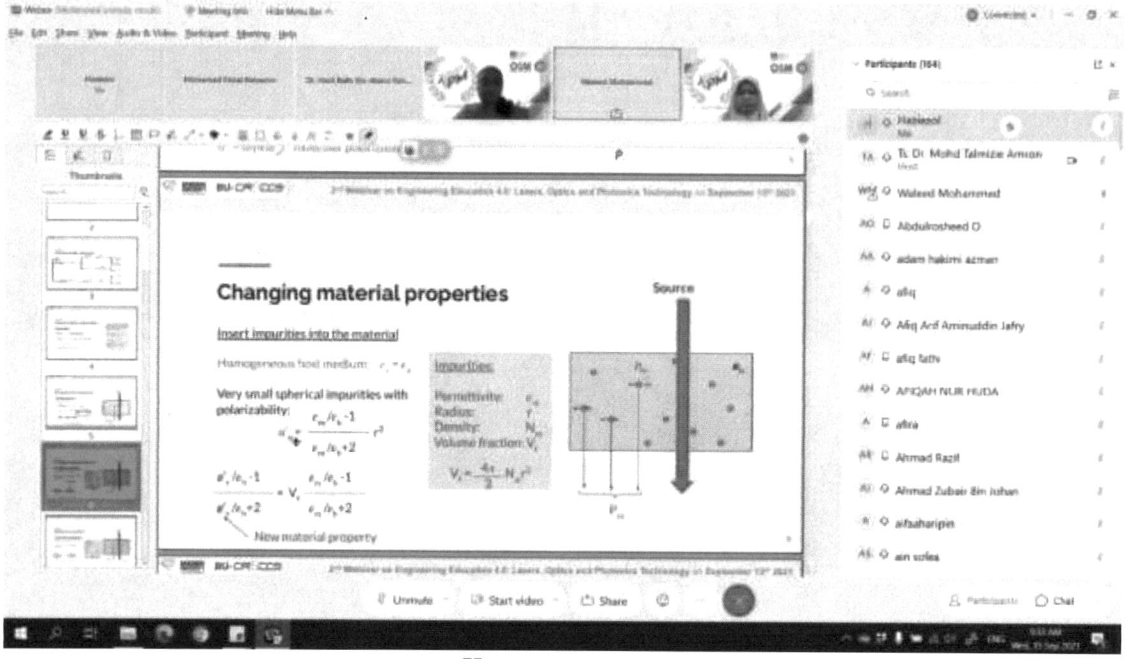

Keynote speaker 1

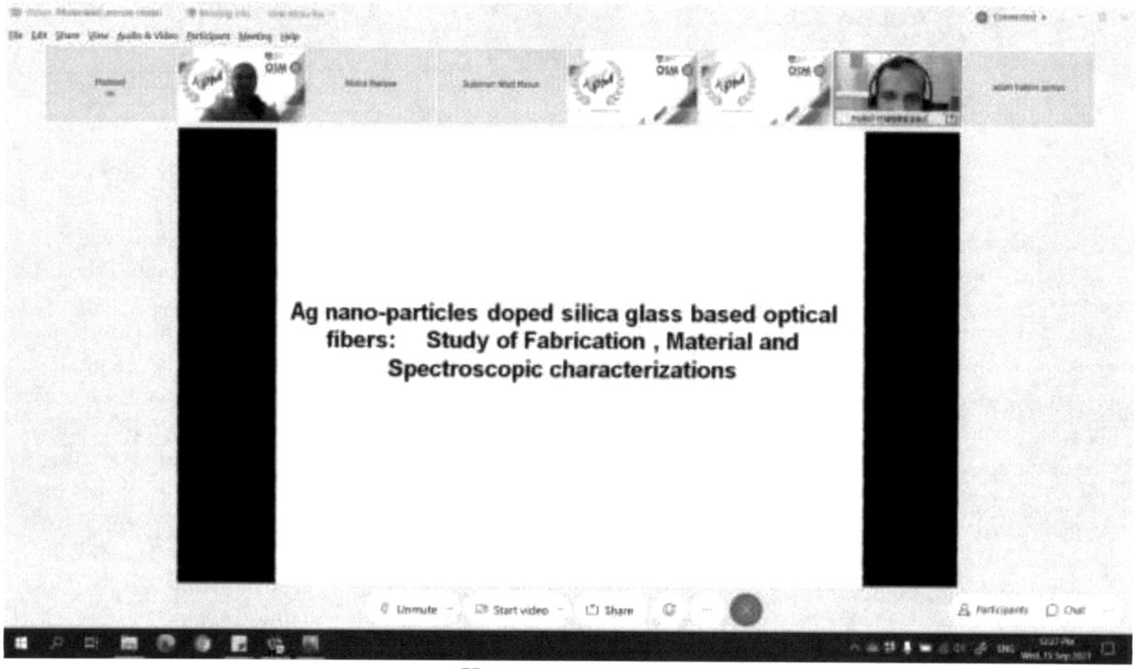

Keynote speaker 2

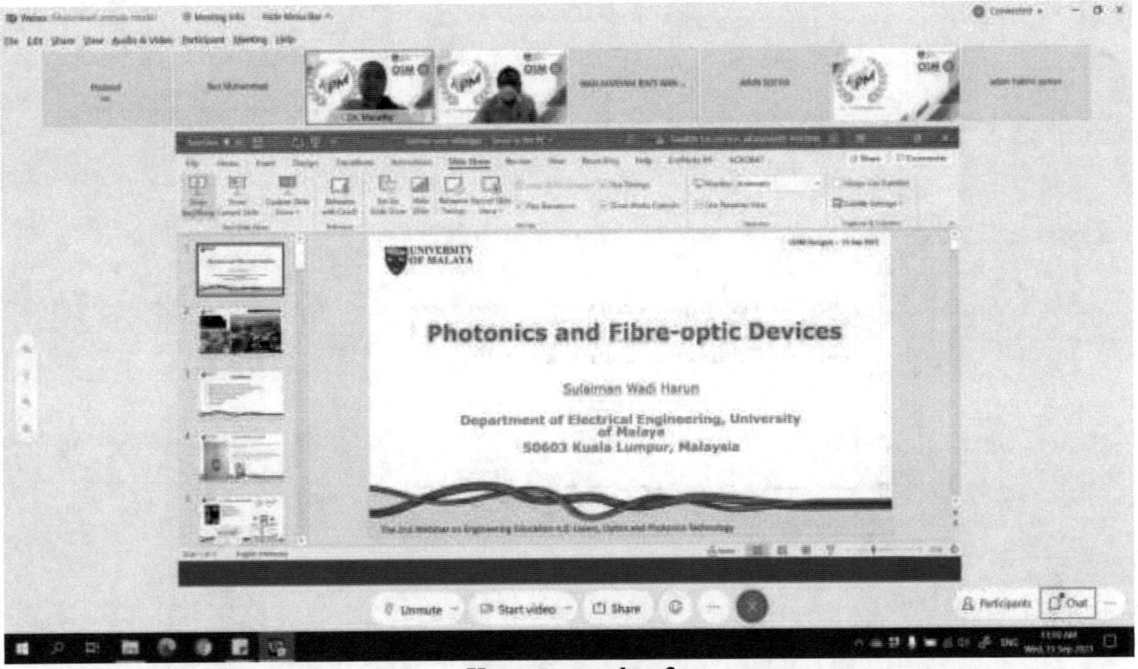

Keynote speaker 3

Participants Day 1

Day 2

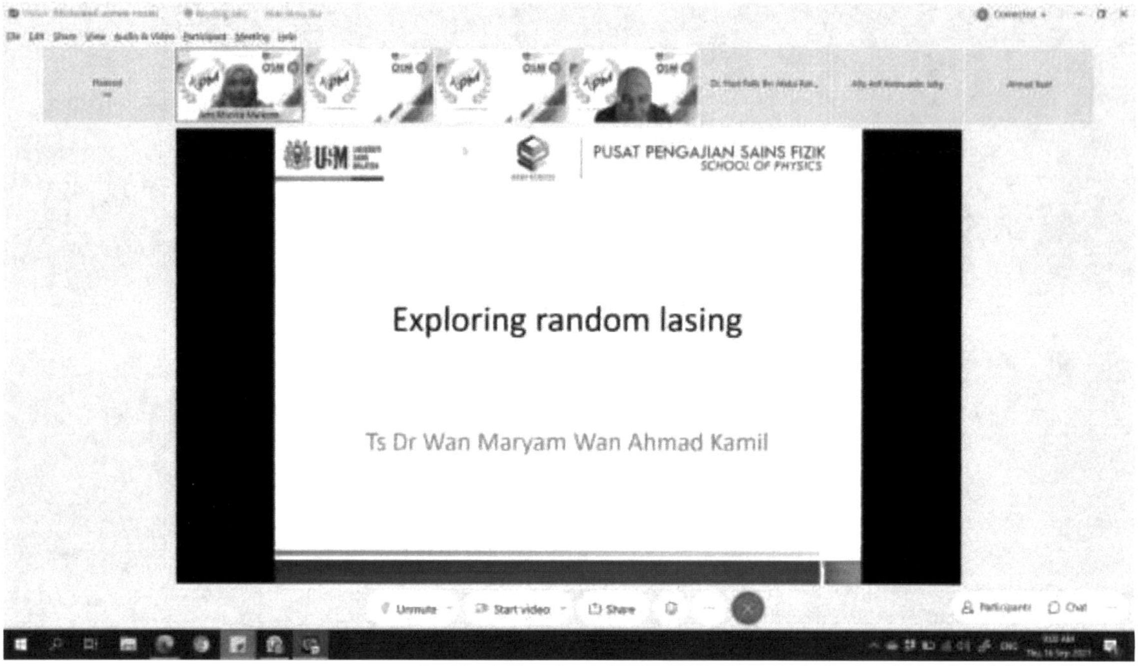

Invited speaker session 1

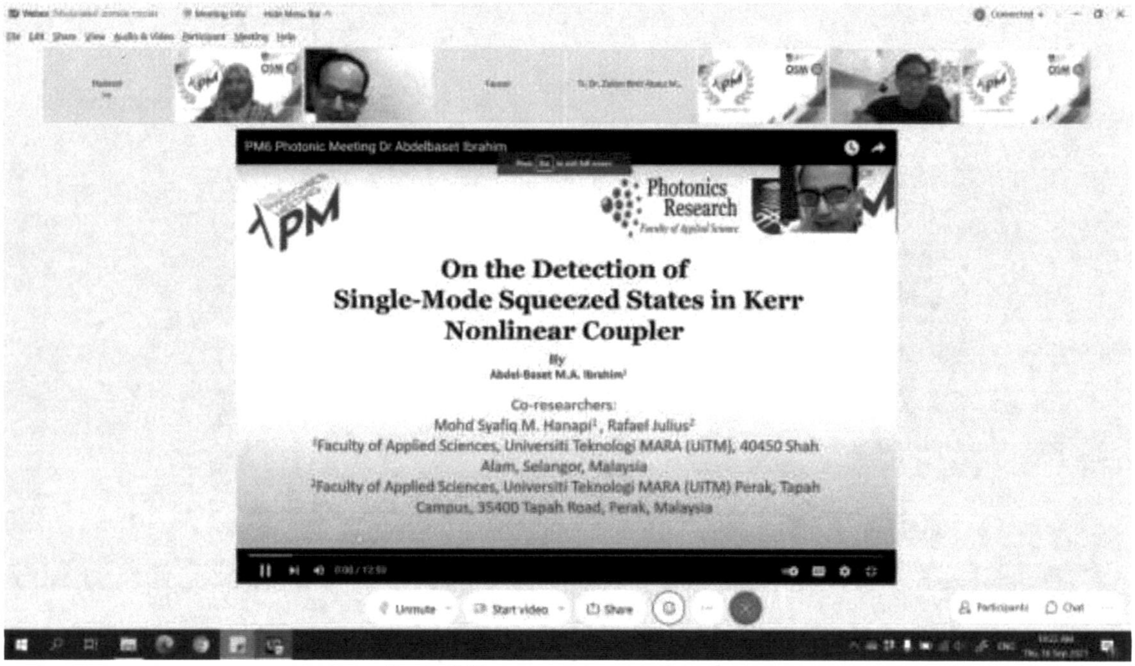

One of the presenter session 1

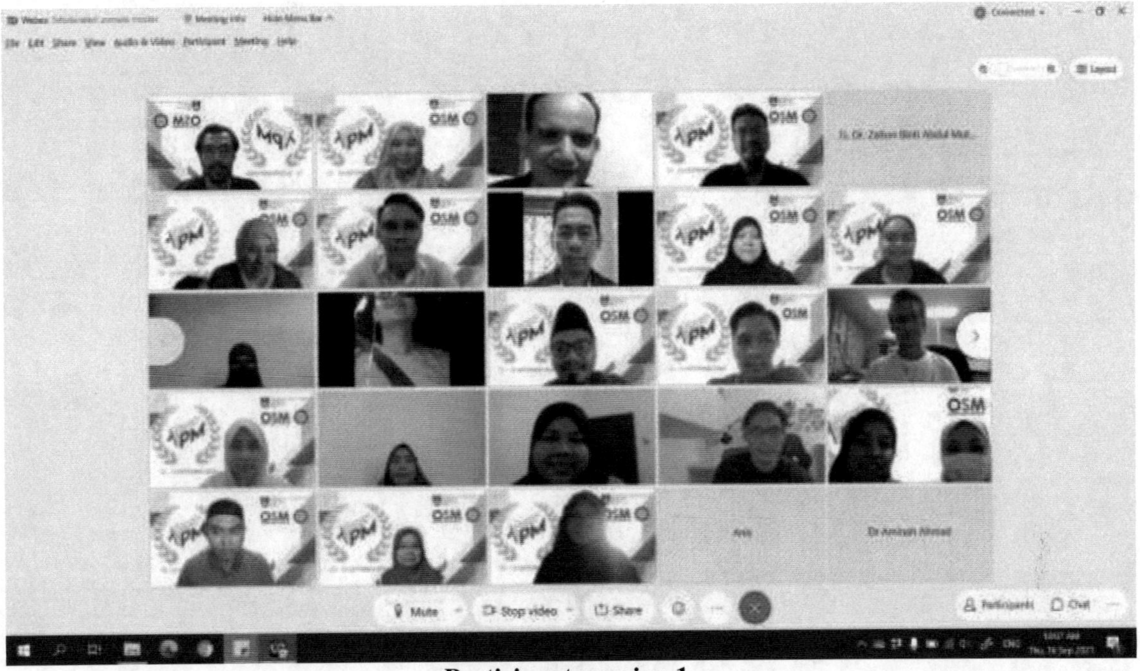

Participants session 1

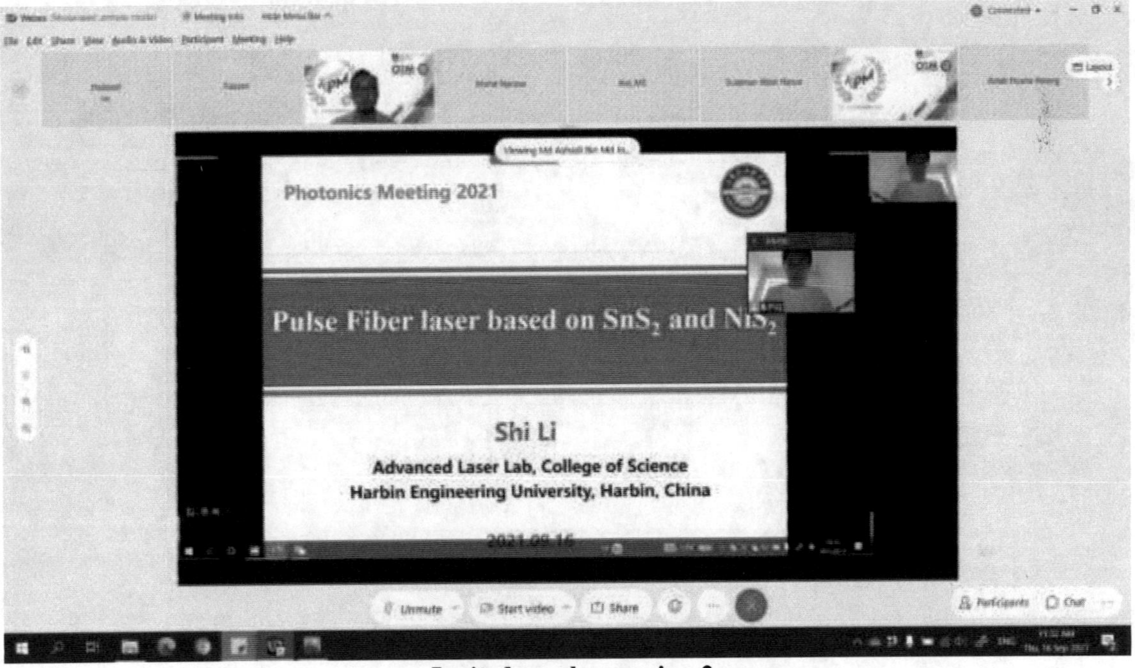

Invited speaker session 2

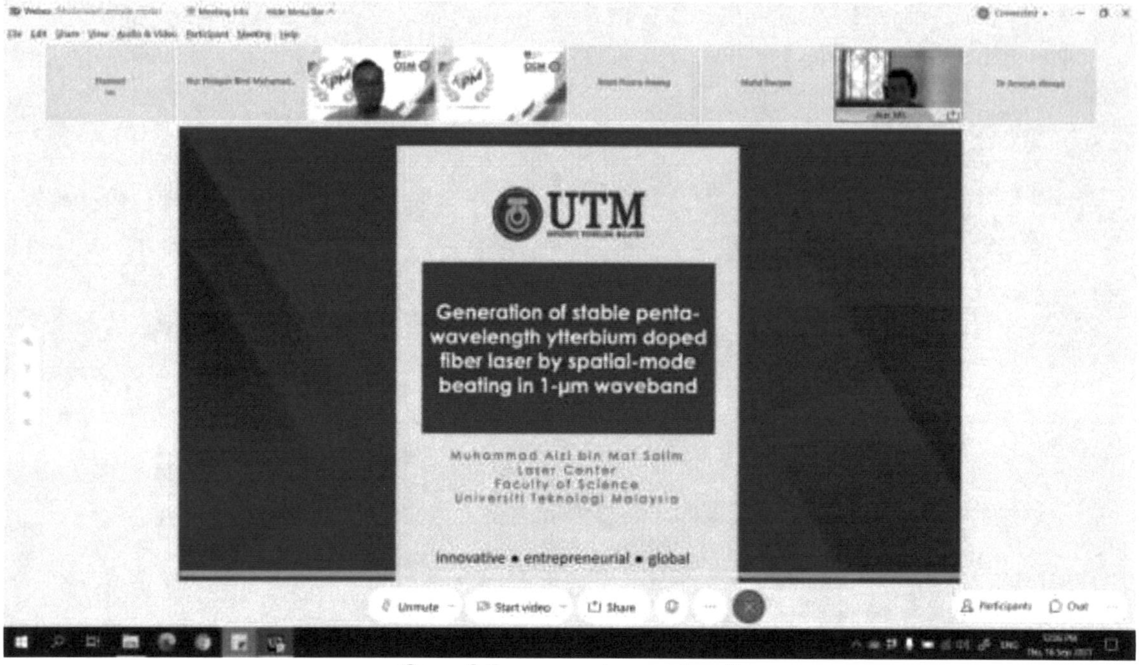

One of the presenter session 2

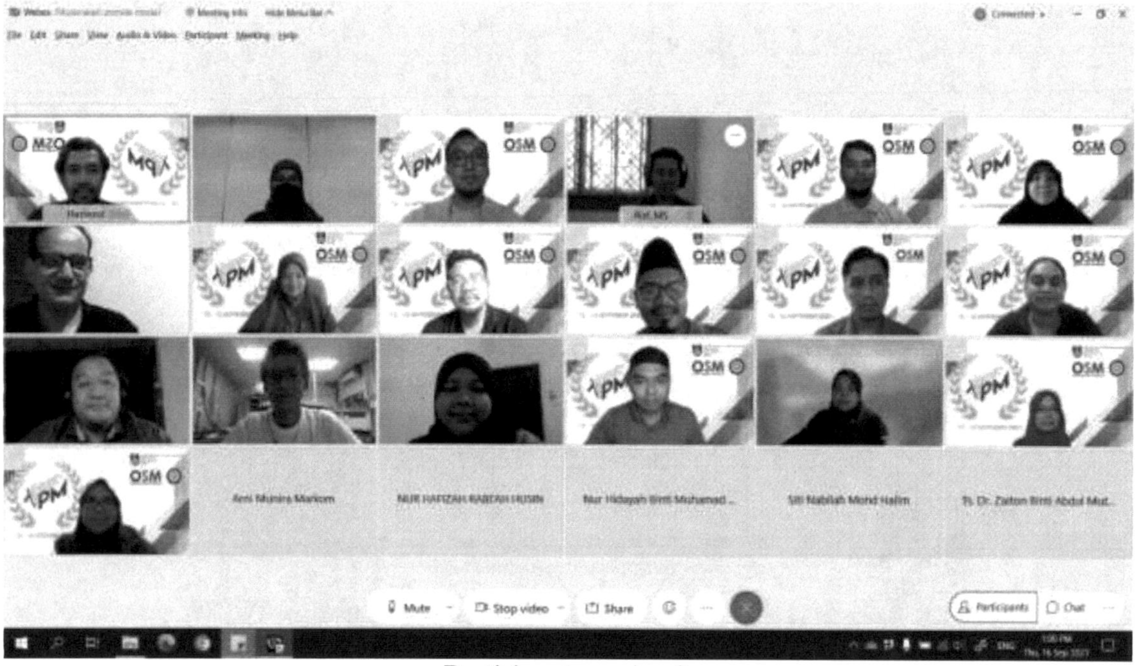

Participants session 2

Journal of Physics: Conference Series **2075** (2021) 011001 doi:10.1088/1742-6596/2075/1/011001

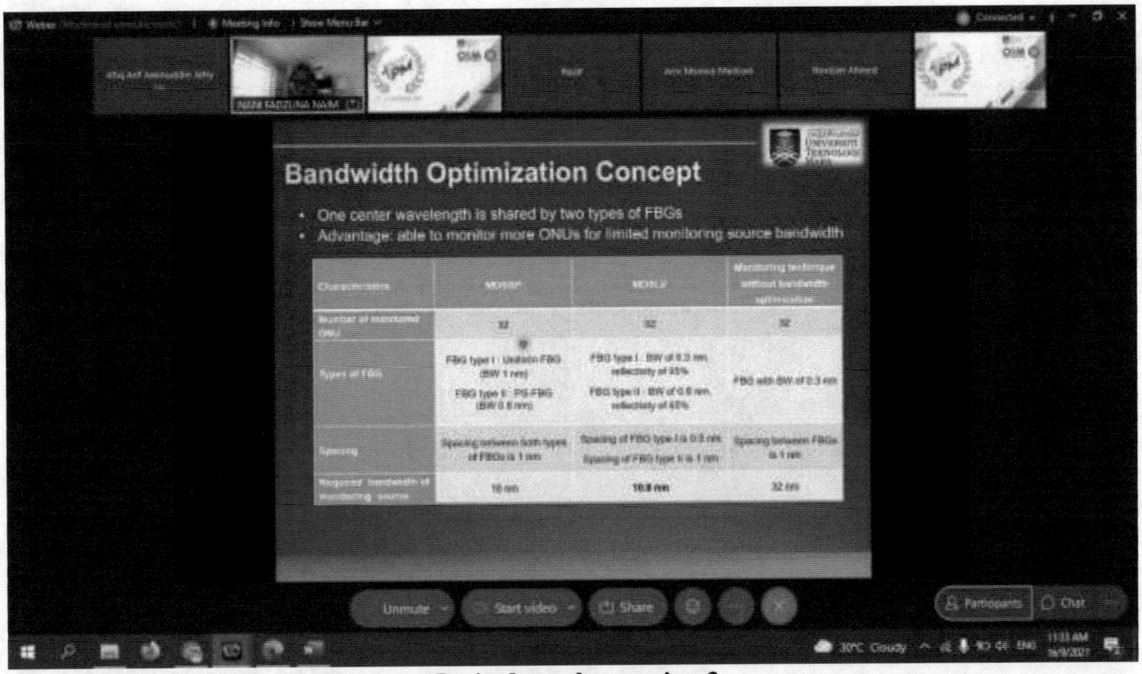

Invited speaker session 3

One of the presenter session 3

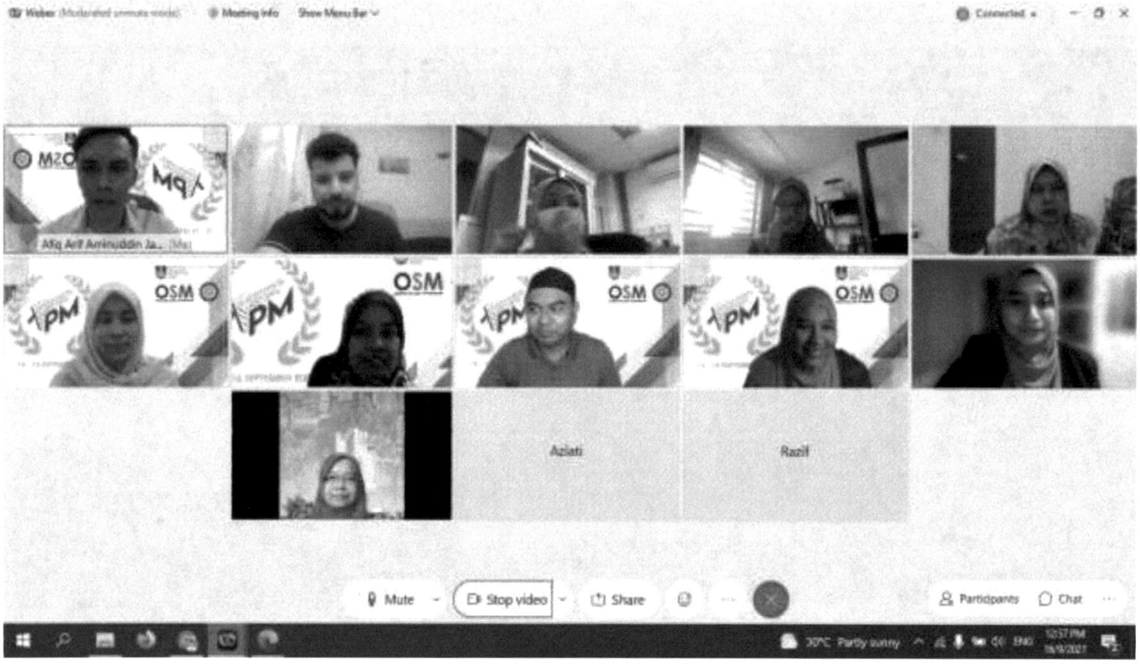

Participants session 3

Invited speaker session 4

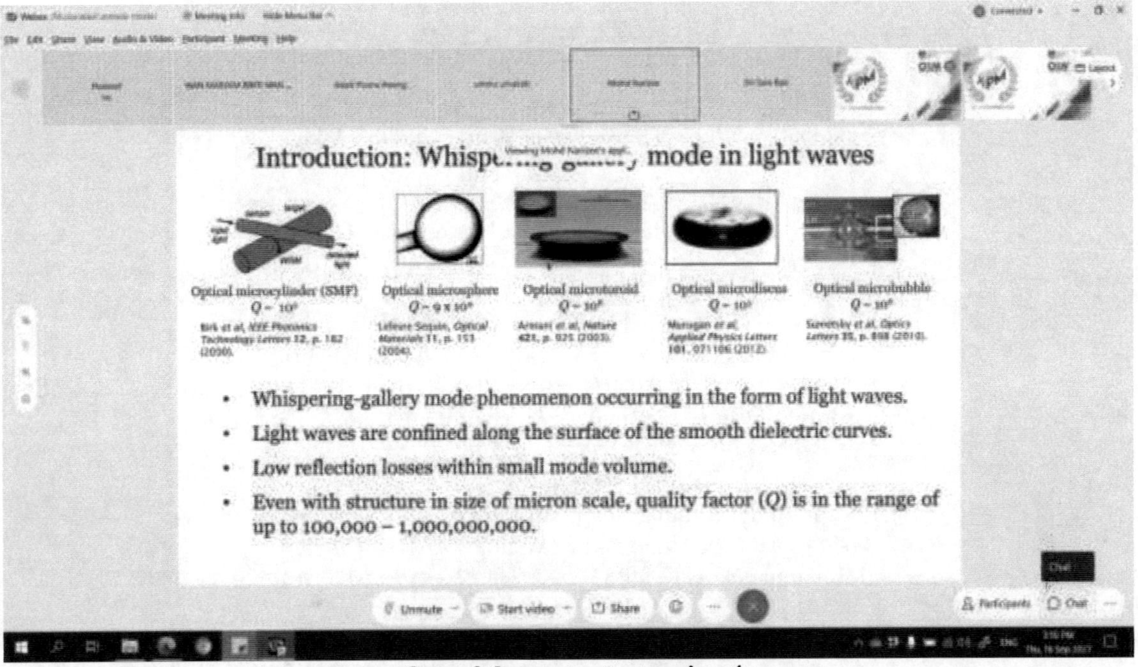

One of the presenter session 4

Participants session 4

List of Committee

Advisor
Sulaiman Wadi Harun, Malaysia (University of Malaya)

Conference Chair
Zulzilawati Jusoh (Universiti Teknologi MARA, Dungun)

Secretary
Nur Farhanah Zulkipli (University of Malaya)
Suziana Omar (Universiti Teknologi MARA, Dungun)
Ku Siti Syahidah Ku Mohd Noh (Universiti Teknologi MARA, Dungun)
Nur Hafizah Rabi'ah Husin (Universiti Teknologi MARA, Dungun)

Bursary
Hazli Rafis Abdul Rahim (Universiti Teknikal Malaysia Melaka)
Norizan Ahmed (Universiti Teknologi MARA, Dungun)
Muhammad Farid Mohd Rusdi (University of Malaya)

Registration
Hasnorhafiza Husni (Universiti Teknologi MARA, Dengkil)
Md Ashadi Md Johari (Universiti Teknikal Malaysia Melaka)
Aminah Ahmad (Universiti Teknikal Malaysia Melaka)
Nur Azwani Mohamad Azmin (Universiti Teknologi MARA, Dungun)
Suhaily Maizan Abdul Manaf (Universiti Teknologi MARA, Dungun)

Publication & Technical Program
Mohd Hafiz Jali (Universiti Teknikal Malaysia Melaka)
Mohamad Faizal Baharom (Universiti Teknikal Malaysia Melaka)
Arni Munira Markom (Universiti Teknologi MARA, Johor)
Anas Abdul Latiff (Universiti Teknikal Malaysia Melaka)
Wan Maryam Wan Ahmad Kamil (Universiti Sains Malaysia)
Mohd Fauzi Ab Rahman (Universiti Teknikal Malaysia Melaka)

Publicy & Marketing
Ahmad Razif Muhammad (Universiti Kebangsaan Malaysia)
Siti Sara Rais (Universiti Teknologi MARA, Dungun)
Nani Fadzlina Binti Naim (Universiti Teknologi MARA, Shah Alam)

Program Coordinator & Protocol
Malathy Batumalay (INTI International University)
Zuraidah Harith (INTI International University)
Aziati Husna Awang (Universiti Teknologi MARA, Shah Alam)
Mohd Talmizie Amron (Universiti Teknologi MARA, Kuala Terengganu)
Haziezol Helmi Mohd Yusof (Universiti Teknikal Malaysia Melaka)
Afiq Arif Aminuddin Jafry (Taylor's College Lakeside Campus)

Peer review declaration

All papers published in this volume of Journal of Physics: Conference Series have been peer reviewed through processes administered by the Editors. Reviews were conducted by expert referees to the professional and scientific standards expected of a proceedings journal published by IOP Publishing.

- **Type of peer review: Single-blind**
- **Conference submission management system: Conference Management Toolkit (CMT)**
- **Number of submissions received: 23**
- **Number of submissions sent for review: 23**
- **Number of submissions accepted: 21**
- **Acceptance Rate (Number of Submissions Accepted / Number of Submissions Received X 100): 91.3%**
- **Average number of reviews per paper: 2**
- **Total number of reviewers involved: 15**
- **Any additional info on review process: -**
- **Contact person for queries:**
Name : Dr. Mohd Hafiz Bin Jali
Affiliation: Unversiti Teknikal Malaysia Melaka
Email : mohd.hafiz@utem.edu.my

Squeezed states generation in an array of Linear and Nonlinear Waveguides

R Julius[1,2*], **A-B M A Ibrahim**[2], **A N Alias**[1] **and M S A Halim**[3]

[1] Faculty of Applied Sciences, Universiti Teknologi MARA (UiTM) Perak, Tapah Campus, 35400 Tapah Road, Perak, Malaysia
[2] Faculty of Applied Sciences, Universiti Teknologi MARA (UiTM), 40450 Shah Alam, Selangor, Malaysia
[3] Faculty of Computer and Mathematical Sciences, Universiti Teknologi MARA (UiTM) Perak, Tapah Campus, 35400 Tapah Road, Perak, Malaysia

[*] Corresponding author's email: rafaeljulius@uitm.edu.my

Abstract. We demonstrate the generation of squeezed states of light due to the second harmonic generation and Kerr effect in an array of nonlinear waveguides mediated through a linear one. We characterized the electromagnetic field by a quantum mechanical Hamiltonian and the density operator time evolution is obtained from the Von-Neumann equation of motion. Using the quasiprobability positive P of phase space representation, the classical Fokker-Planck equation is obtained from the master equation and translated to its classical matching set of nonlinear differential equations. We showed that because of the new possibilities of correlation between the linear and nonlinear channel waveguides, highly nonclassical light may be produced.

1. Introduction

To take advantage of quantum technology, it is vital to produce and regulate quantum systems and devices. In this context, guided wave structures that entail the manipulation of light and its interaction with nonlinear media are one of the most well-known systems. The generation and control of highly nonclassical light in an array of guided wave structures can be realized using various configurations such as the manipulation of several modes [1], waveguides [2], or the nonlinear processes of the interactive media such as second harmonic generation (SHG) [3, 4] and Kerr effects [5]. SHG may have fewer absorptive losses and reduce quantum noise more effectively [6]. On the other hand, Kerr effects are a great mechanism for generating quantum states of light since they do not require phase matching and the nonlinear response is very fast [7]. Coupled waveguides arrays without nonlinear effects may not be able to generate nonclassical light on their own, however, they might be utilized for transport and quantum walk of nonclassical light [8-11].

As many novel physical dynamics can emerge from a system with combined nonlinearities [12], it's relevant to look into the potential of generating highly nonclassical light in such a system. This paper investigates the generation of squeezed states in an array of waveguides with both SHG and third-order Kerr nonlinearity mediated by a linear one. When a coherent input field interacts with these nonlinear media, fascinating quantum phenomena such as squeezing can be produced. In this system, the electromagnetic field is characterized by a quantum mechanical Hamiltonian, whereas its propagation is given by the master equation for the density matrix. Using the positive P representation of the phase space [13, 14], the master equation is translated to its classical matching set of nonlinear differential equations, and adequate descriptions of the quantum states may be obtained.

2. Theoretical Aspect

As illustrated in Figure 1, we consider the generation of squeezed states in an array of waveguides due to the SHG and Kerr effect.

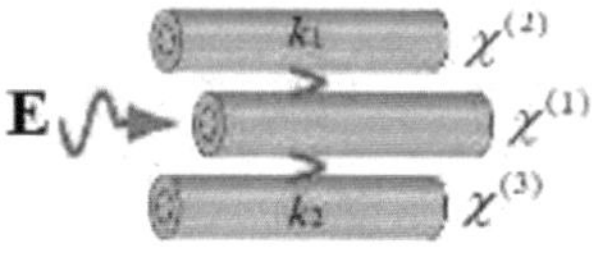

Figure 1. Schematic illustration of the model

The channel waveguides are set near together to allow the evanescent fields to overlap, allowing the coupled modes to exchange energy regularly during wave propagation. The model under consideration can be described by the following effective Hamiltonian,

$$\hat{H} = \frac{h}{2\pi}\Big\{\omega\left(\hat{a}^{\dagger}\hat{a} + \hat{b}^{\dagger}\hat{b} + \hat{A}^{\dagger}\hat{A}\right) + \Omega\hat{B}^{\dagger}\hat{B} + i\frac{\chi}{2}\hat{A}^{\dagger 2}\hat{B} - \hat{A}^{2}\hat{B}^{\dagger} + g\hat{b}^{\dagger 2}\hat{b}^{2} + k_1\left(\hat{a}\hat{A}^{\dagger} + \hat{A}\hat{a}^{\dagger}\right) + k_2\left(\hat{a}\hat{b}^{\dagger} + \hat{b}\hat{a}^{\dagger}\right)\Big\} \tag{1}$$

Herein, h defines Planck constant, ω and Ω are the fundamental (FH) and second harmonic (SH) frequencies of the input field, χ and g are the nonlinear coefficients proportional to the quadratic $\chi^{(2)}$ and cubic $\chi^{(3)}$ nonlinearities and k_i (i = 1, 2) is the evanescent linear coupling between the waveguides. The components $\hat{a}^{\dagger}\hat{a}, \hat{b}^{\dagger}\hat{b}, \hat{A}^{\dagger}\hat{A}$ and $\hat{B}^{\dagger}\hat{B}$ are the bosonic operators for the FH ($\hat{a}^{\dagger}\hat{a}, \hat{b}^{\dagger}\hat{b}, \hat{A}^{\dagger}\hat{A}$) and SH ($\hat{B}^{\dagger}\hat{B}$) modes respectively. Appropriate representations of quantum states can be obtained by applying the phase space representation to the current system [15]. In this case, the quantum operator Von-Neumann equation of the density operator must be converted to the corresponding classical Fokker-Planck (FP) equation using positive P representation. In turn, the classical FP equation of the quasiprobability distribution can be mapped to a set of stochastic partial differential equations using Ito Calculus rules. This yield the following set of:

$$\begin{aligned}
\dot{\alpha} &= -i\left(\omega\alpha + ik_1\varsigma + ik_2\beta\right)\\
-\dot{\alpha}^{*} &= -i\left(\omega\alpha^{*} + ik_1\varsigma^{*} + ik_2\beta^{*}\right)\\
\dot{\beta} &= -i\left(\omega\beta + 2g\beta^{*}\beta^{2} + ik_2\alpha\right) + \sqrt{-2ig}\,\beta\eta_1(z)\\
-\dot{\beta}^{*} &= -i\left(\omega\beta^{*} + 2g\beta\beta^{*2} + ik_2\alpha^{*}\right) - \sqrt{2ig}\,\beta^{*}\eta_2(z)\\
\dot{\varsigma} &= -i\omega\varsigma + \chi\varsigma^{*}\xi - ik_1\alpha + \sqrt{\chi\xi}\,\eta_3(z)\\
\dot{\varsigma}^{*} &= i\omega\varsigma^{*} + \chi\xi^{*}\varsigma + ik_1\alpha^{*} + \sqrt{\chi\xi^{*}}\,\eta_4(z)\\
\dot{\xi} &= -i\Omega\xi - \frac{1}{2}\chi\varsigma^{2}\\
\dot{\xi}^{*} &= i\Omega\xi^{*} - \frac{1}{2}\chi\varsigma^{*2}
\end{aligned} \tag{2}$$

The overdot in equation (2) reflects the derivatives with respect to the spatial distance z, whereas $\{\alpha, \alpha^{*}\}$, $\{\beta, \beta^{*}\}$, $\{\varsigma, \varsigma^{*}\}$ and $\{\xi, \xi^{*}\}$ are independent stochastic variables corresponding to the operators $\hat{a}^{\dagger}\hat{a}, \hat{b}^{\dagger}\hat{b}, \hat{A}^{\dagger}\hat{A}$ and $\hat{B}^{\dagger}\hat{B}$. The Gaussian noise η_a satisfies the correlation $\eta_i(z) = 0$ and $\eta_i(z)\,\eta_j(z') = \delta_{ij}\delta(z - z')$. Squeezing is numerically examined in terms of the propagating field quadrature variances, given by

$$S_X = 4\langle(\Delta\hat{X})^2\rangle - 1 \leq 0 \quad ; \quad S_Y = 4\langle(\Delta\hat{Y})^2\rangle - 1 \leq 0 \tag{3}$$

where

$$2\hat{X} = (\hat{O} + \hat{O}^{\dagger}) \quad ; \quad 2i\hat{Y} = (\hat{O} - \hat{O}^{\dagger}) \tag{4}$$

$$\left\langle(\Delta X)^2\right\rangle = \frac{1}{4}\left[\left\langle O^2\right\rangle + \left\langle O^{*2}\right\rangle + 2\left\langle O^* O\right\rangle + 1 - \left\langle O\right\rangle^2 - \left\langle O^*\right\rangle^2 - 2\left\langle O\right\rangle\left\langle O^*\right\rangle\right]$$
$$\left\langle(\Delta Y)^2\right\rangle = \frac{1}{4}\left[-\left\langle O^2\right\rangle - \left\langle O^{*2}\right\rangle + 2\left\langle O^* O\right\rangle + 1 + \left\langle O\right\rangle^2 + \left\langle O^*\right\rangle^2 - 2\left\langle O\right\rangle\left\langle O^*\right\rangle\right] \quad (5)$$

and $[\hat{O}_i, \hat{O}_j^\dagger] = \delta_{ij}, \hat{O}\hat{O}^\dagger = \hat{a}^\dagger\hat{a}, \hat{b}^\dagger\hat{b}, \hat{A}^\dagger\hat{A}$. For convenience in numerical simulation, the input parameters have been cast to dimensionless form and are fixed at $\chi = 0.01$, $g = 10^{-7}$, $\omega = 2$ and $\Omega = 2\omega$ with coherent state initialization in all channels.

3. Discussion

Squeezing reduces the intrinsic quantum noise in the field quadrature to below the standard quantum limit. This allows for more exact measurements than would otherwise be achievable, and it has the potential to advance photonic quantum computing [16, 17]. Using the interpolation data of peak amplitude squeezing and antisqueezing, figure 2 depicts the generation of squeezed states in the $\chi^{(1)}$ channel at various k_i states where $\chi^{(1)}$ is a coefficient proportional to the linear susceptibility of the material. For three channels interaction, the $\chi^{(1)}$ channel has a linear coupling constant $k_1 = k_2 = 1$, whereas for $k_1 = 1$, $k_2 = 0$ and $k_1 = 0$, $k_2 = 1$ correspondingly, the linear coupling is simplified to $\chi^{(1)}$- $\chi^{(2)}$ and $\chi^{(1)}$- $\chi^{(3)}$ interaction. Thus, as a result of quantum correlations between the field modes, the system produces a periodic squeezed state. Within the parameter ranges considered in this study, squeezing due to the SHG provides the least amount of squeezing in the $\chi^{(1)}$ channel. In early evolution, the quadrature variances for $\chi^{(1)}$- $\chi^{(3)}$ interaction exhibits the strongest squeezing. Compared to the $\chi^{(1)}$- $\chi^{(3)}$ interaction, squeezing owing to both the SHG and Kerr effects ($k_1 = k_2 = 1$) is particularly weaker at this point, however, when squeezing grew in proportion to the interaction length, the $\chi^{(1)}$- $\chi^{(2)}$- $\chi^{(3)}$ interaction demonstrated the largest amount of squeezing.

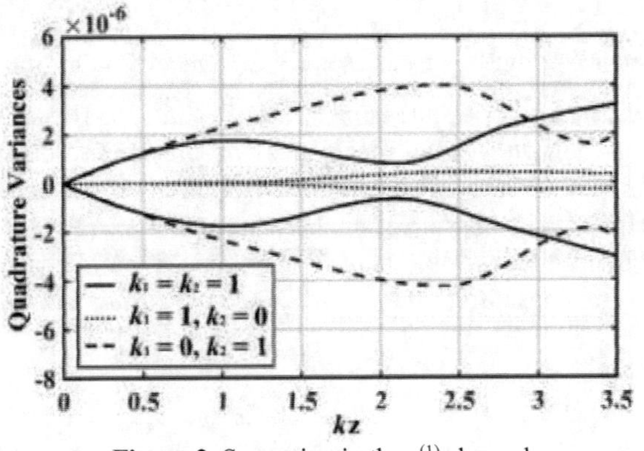

Figure 2. Squeezing in the $\chi^{(1)}$ channel

Figure 3 shows the generation of squeezed states in the $\chi^{(2)}$ channel using the interpolation data of peak amplitude squeezing and antisqueezing following equation (3). While the SHG and Kerr nonlinearity can improve squeezing in the $\chi^{(1)}$ channel, the presence of Kerr nonlinearity mediated by the $\chi^{(1)}$ channel limited the maximal squeezing of the FH field achievable in the $\chi^{(2)}$ channel on an individual mode basis. When only the $\chi^{(2)}$ nonlinearity is addressed, maximal squeezing can be generated. Furthermore, squeezed light at higher frequency ranges may be advantageous for applications at different wavelengths. We found that, it is possible to extend squeezing over the FH frequency. Nonetheless, when compared to squeezing at the FH frequency, squeezing at the SH frequency is much weaker, with a tendency for collapse and revival pattern in the quadrature variance evolution. The generation of squeezed states in the $\chi^{(2)}$ channel at SH frequency is shown in figure 4.

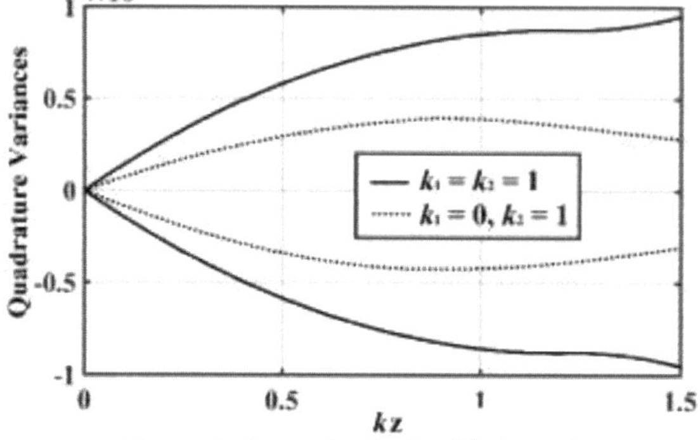

Figure 3. Squeezing in the $\chi^{(2)}$ channel

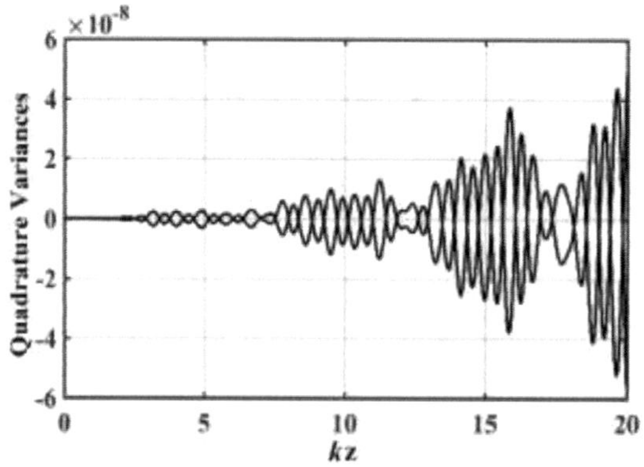

Figure 4. Squeezing at the second harmonic frequency

Furthermore, figure 5 shows single-mode squeezing in the $\chi^{(3)}$ channel. These results resemble those obtained in $\chi^{(2)}$ channel, however with varied magnitude and phase. The presence of second-order nonlinearity mediated by the $\chi^{(1)}$ channel might reduce the amount of squeezing in the $\chi^{(3)}$ channel. Squeezing might be better for $\chi^{(1)}$- $\chi^{(3)}$ consideration, however, the amount is limited. The propagating field in the $\chi^{(3)}$ channel is squeezed to a greater extent when there is no interaction at all between the $\chi^{(1)}$ and $\chi^{(3)}$ channels. Therefore, considering the $\chi^{(1)}$-$\chi^{(2)}$-$\chi^{(3)}$ interaction might not be ideal to improve squeezing in this channel.

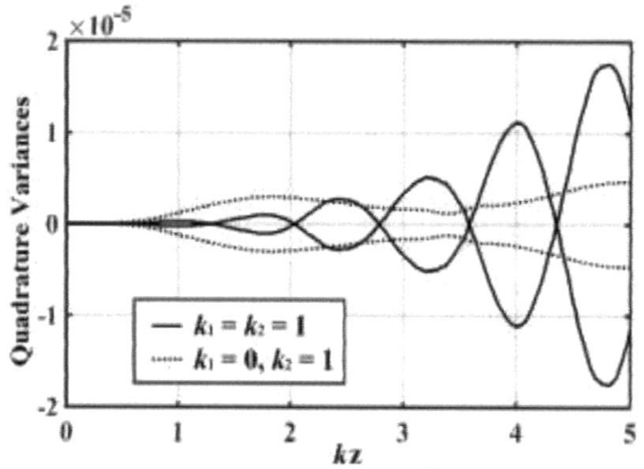

Figure 5. Squeezing in the $\chi^{(3)}$ channel

4. Conclusion

In conclusion, we have investigated the possibility of generating enhanced squeezed states of light in linear waveguide through the combination of SHG and Kerr nonlinearity. The presence of Kerr nonlinearity mediated by the $\chi^{(1)}$ channel limited the maximal squeezing achievable at FH frequency in the $\chi^{(2)}$ channel, however, it is possible to extend squeezing at the SH frequency. The current multichannel system combines both linear and nonlinear waveguides is versatile and potentially useful as squeezed light generator especially in the linear media.

References

[1] Julius R, Ibrahim A-B and Deni M 2014 Quantum state generation in a four-mode Kerr nonlinear directional coupler *Laser Physics* **24** 035202

[2] Julius R, Ibrahim A-B M A, Choudhury P K and Eleuch H 2018 On the nonclassical dynamics of cavity-assisted four-channel nonlinear coupler *Chinese Phys.* B **27** 114206

[3] Julius R, Ibrahim A-B M A, Choudhury P K and Eleuch H 2019 Quantum states generation in multichannel directional coupler with second-order nonlinearity *Optik* **186** 212–20

[4] Julius R, Ibrahim A-B M A, Eleuch H and Choudhury P K 2019 Sub-Poissonian photon squeezing and entanglement in optical chain second harmonic generation *Journal of Modern Optics* **66** 1129–38

[5] Julius R, Ibrahim A-B M A, Abd-Rahman M K and Choudhury P K 2018 Quantum dynamics of a four-channel Kerr nonlinear directional coupler *Opt Rev* **25** 563–70

[6] Li R-D and Kumar P 1994 Quantum-noise reduction in traveling-wave second-harmonic generation *Phys. Rev.* A **49** 2157–66

[7] Ibrahim A-B, Umarov B and Wahiddin R 2000 Squeezing in the Kerr nonlinear coupler via phase-space representation *Phys. Rev.* A **61** 43804

[8] Tang H, Lin X-F, Feng Z, Chen J-Y, Gao J, Sun K, Wang C-Y, Lai P-C, Xu X-Y, Wang Y, Qiao L-F, Yang A-L and Jin X-M 2018 Experimental two-dimensional quantum walk on a photonic chip *Science Advances* **4** eaat3174

[9] Peruzzo A, Lobino M, Matthews J C F, Matsuda N, Politi A, Poulios K, Zhou X-Q, Lahini Y, Ismail N, Wörhoff K, Bromberg Y, Silberberg Y, Thompson M G and OBrien J L 2010 Quantum Walks of Correlated Photons *Science* **329** 1500–3

[10] Rai A, Agarwal G S and Perk J H H 2008 Transport and quantum walk of nonclassical light in coupled waveguides *Phys. Rev.* A **78** 042304

[11] Rodríguez-Lara B M 2014 Propagation of non-classical states of light through one-dimensional photonic lattices *J. Opt. Soc. Am.* **B** 31 878

[12] Zhu E, Zhao C and Li H 2020 Frequency-domain model of optical frequency-comb generation in optical resonators with second- and third-order nonlinearities *Phys. Rev.* A **102** 053508

[13] Gardiner C, Gilchrist A and Drummond P 1993 Using the positive P-representation Fundamentals of Quantum Optics III Lecture Notes in Physics ed F Ehlotzky (Berlin, Heidelberg: Springer) pp 239–53

[14] Gilchrist A, Gardiner C and Drummond P 1997 Positive P representation: Application and validity *Phys. Rev.* A **55** 3014

[15] Olsen M and Bradley A 2009 Numerical representation of quantum states in the positive-P and Wigner representations Optics Communications **282** 3924–9

[16] Fukui K, Tomita A, Okamoto A and Fujii K 2018 High-Threshold Fault-Tolerant Quantum Computation with Analog Quantum Error Correction *Phys. Rev.* **X** 8 021054

[17] Takeda S and Furusawa A 2019 Perspective: Toward large-scale fault-tolerant universal photonic quantum computing *APL Photonics* **4** 060902

Acknowledgment

This work is supported by the Geran Khas Insentif Penyelidikan Perak (GKIPP) with grant number 900-KPK/PJI/GKIPP/01(0010/2020) received from the Universiti Teknologi MARA, Perak Branch, Malaysia.

On the Detection of Single-Mode Squeezed States in Kerr Nonlinear Coupler

Mohd Syafiq M. Hanapi[1], **Abdel-Baset M.A. Ibrahim**[1*], **Rafael Julius**[2]

[1] Faculty of Applied Sciences, Universiti Teknologi MARA (UiTM), 40450 Shah Alam, Selangor, Malaysia
[2] Faculty of Applied Sciences, Universiti Teknologi MARA (UiTM) Perak, Tapah Campus, 35400 Tapah Road, Perak, Malaysia

[*] Corresponding author's email: abdelbaset@uitm.edu.my

Abstract. The generation of squeezed states of light in two guided waves Kerr nonlinear coupler (KNLC) was examined using both the analytical perturbative (AP) and the short-time approximation (STA) method. A comparative analysis between these two methods is provided. We have found that, at certain combinations of input parameters, the STA method may not be able to detect the generation of squeezed states of light in the current KNLC system. Consequently, some essential physics could be lost. On the other hand, for the AP method, all time-dependent terms are included in the mode solutions which improves its sensitivity to detect the generation of squeezed states.

1. Introduction

Squeezed states of light or simply squeezed light has special properties that allow it to surpass the standard quantum limit. Squeezed light is crucial in future technological applications. It can mainly be applied to enhance the sensitivity of optical measuring equipment [1], [2]. Furthermore, squeezed light has the potential to initiate the quantum teleportation phenomenon [3], [4]. Short-range teleportation is essential to deliver quantum computational data [5]. Hence, squeezed light is important to put the quantum computer into action. It is also important in many other applications such as communication [6]–[8], gravitational wave detection [9], [10], quantum imaging [11]–[13], and quantum error correction coding [14], [15]. Due to all these futuristic applications, both theoretical and experimental research on squeezed light was given great attention [16]–[20].

Directional coupler essentially consists of two waveguides placed closely along each other to allow the propagating field-modes to linearly couple via evanescent waves. A schematic diagram of this devices is shown in Fig. 1. The strength of linear coupling is inversely proportional to the separation between the channels. In Kerr nonlinear coupler, the propagating optical field in the waveguides produces an optical Kerr effect due to the nonlinear interaction with the guiding medium. Recently, Kerr nonlinear coupler (KNLC) became a versatile source of quantum phenomena [21]–[25]. This is because it is a simple-structured device and is easily integrated with/within other devices such as photonic chips, quantum circuits, and optical logic gates [26]–[29].

A standard method to investigate squeezed state generation in KNLC is the short-time approximation (STA) method where an analytical solution is assumed in the form of second-order Taylor series where the higher-order terms are ignored. This approximation leads to the loss of some essential physics in the short-time approximation method as will be evident in this work. Alternatively, to examine the generation of squeezed states in KNLC, we present an analytical solution to the coupled Heisenberg equations of motion for the propagating modes in the form of Baker–Hausdorff (BH) formula. Based on BH formula, an intuitive solution of the modes could be obtained with fully deterministic time-dependent coefficients. In this method, the nonlinearity in the system is assumed to be small enough to be treated as a small perturbation, hence the name *analytical-perturbative* or simply the (AP) method. Previously, the AP method was employed to examine the quantum properties generated in mechanisms other than KNLC such as optomechanical systems [30], [31], light-semiconductor interaction [32], Raman process [33]–[36], four-wave mixing [37], and Bose-Einstein Condensation [38]–[40].

In this work, we aim at investigating the generation of squeezed states of light in KNLC using the AP method. In particular, we aim at comparing both methods and examine their efficiency in detecting single-mode squeezed states generated in Kerr nonlinear coupler. This comparison could provide useful physical insights on the accuracy, capability, and reliability of these two methods in detecting the nonclassical effects in quantum optical devices. In the following sections, we present the basic mathematical formulation of both the STA and AP method, a discussion of the main results, and finally, a conclusion is drawn.

2. Mathematical formulation

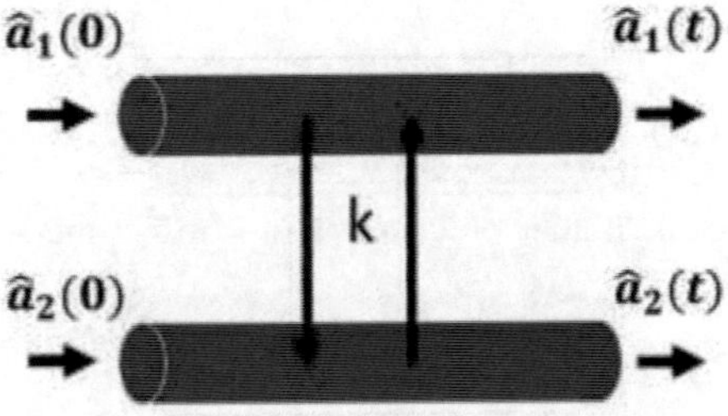

Figure 1. A schematic diagram of the two-channel Kerr coupler. Each waveguide is traversed by one fundamental mode. The light modes are represented by their respective annihilation operators, $\hat{a}_1$ and $\hat{a}_2$ while the quantity k symbolizes the strength of linear coupling.

In this section, the complete mathematical description for KNLC is presented. The two-mode system under consideration can be characterized by the following Hamiltonian operator.

$$\hat{H} = \hbar\left[\omega_1\hat{a}_1^\dagger\hat{a}_1 + \omega_1\hat{a}_2^\dagger\hat{a}_2 + g\left(\hat{a}_1^{\dagger 2}\hat{a}_1^2 + g\hat{a}_2^{\dagger 2}\hat{a}_2^2\right) + k\left(\hat{a}_1^\dagger\hat{a}_2 + \hat{a}_2^\dagger\hat{a}_1\right)\right] \tag{1}$$

In Eq. 1, $\hbar$ is the reduced Planck constant, and $\hat{a}_j^\dagger$ ($\hat{a}_j$) is the creation (annihilation) operator of the modes. The first two terms on the right-side of Eq.1 describe the propagating terms of both modes at a frequency ω_j. The third term represents the nonlinear coupling between the modes and the Kerr medium. This interaction is controlled by the nonlinear coupling coefficient, g . The final term describes the linear coupling interaction between both modes, with linear coupling strength k. The overall energy in the mode interaction is conserved since the energy loss by one mode corresponds to an energy gain by the opposite mode. By substituting Eq. (1) into the general form of Heisenberg equation of motion $\frac{d\hat{a}_j}{dt} = i[\hat{a}_j, \hat{H}]$ we obtained the following.

$$\frac{d\hat{a}_1}{dt} = i\left[\omega_1\hat{a}_1 + 2g\hat{a}_1^\dagger\hat{a}_1^2 + k\hat{a}_2\right] \tag{2}$$

$$\frac{d\hat{a}_2}{dt} = i\left[\omega_2\hat{a}_2 + 2g\hat{a}_2^\dagger\hat{a}_2^2 + k\hat{a}_1\right] \tag{3}$$

Each one of These equations describes the motion of an individual light mode. The system (2)-(3) has the form of coupled nonlinear differential equations, hence analytical closed-form solutions are readily unavailable. Here, we attempt to obtain the closed-form analytical solution for such a system under certain approximations using both the STA and AP methods

2.1. Short-Time Approximation (STA) Method

In this method, solutions for (2)-(3) can be derived using the McLaurin series. This method employs the short-time approximation that neglects terms beyond t^2. Under this approximation, the solutions will take the following form

$$\begin{aligned}\hat{a}_1(\tau) = {} & \hat{a}_1(0) + i\left[\tilde{\omega}_1\hat{a}_1(0) + 2\tilde{g}\hat{a}_1^\dagger(0)\hat{a}_1^2(0) + \tilde{k}\hat{a}_2(0)\right]\tau - \tfrac{1}{2}\left[\tilde{\omega}_1^{\,2}\hat{a}_1(0) + \tilde{\omega}_1\tilde{k}\hat{a}_2(0) + 2\tilde{g}\tilde{k}\hat{a}_2^\dagger(0)\hat{a}_1^2(0)\right. \\ & +4\tilde{\omega}_1\tilde{g}\hat{a}_1^\dagger(0)\hat{a}_1^2(0) + 4\tilde{g}\tilde{k}\hat{a}_1^\dagger(0)\hat{a}_1(0)\hat{a}_2(0) + 4\tilde{g}^2\hat{a}_1^\dagger(0)\hat{a}_1(0)\hat{a}_1^\dagger(0)\hat{a}_1^2(0) + \tilde{\omega}_2\tilde{k}\hat{a}_2(0) \\ & \left. +2\tilde{g}\tilde{k}\hat{a}_2^\dagger(0)\hat{a}_2^2(0) + \tilde{k}^2\hat{a}_1(0)\right]\tau^2\end{aligned} \tag{4}$$

$$\begin{aligned}\hat{a}_2(\tau) = {} & \hat{a}_2(0) + i\left[\tilde{\omega}_2\hat{a}_2(0) + 2\tilde{g}\hat{a}_2^\dagger(0)\hat{a}_2^2(0) + \tilde{k}\hat{a}_1(0)\right]\tau - \tfrac{1}{2}\left[\tilde{\omega}_2^{\,2}\hat{a}_2(0) + \tilde{\omega}_2 k\hat{a}_1(0) + 2\tilde{g}\tilde{k}\hat{a}_1^\dagger(0)\hat{a}_2^2(0)\right. \\ & +4\tilde{\omega}_2\tilde{g}\hat{a}_2^\dagger(0)\hat{a}_2^2(0) + 4\tilde{g}\tilde{k}\hat{a}_2^\dagger(0)\hat{a}_2(0)\hat{a}_1(0) + 4\tilde{g}^2\hat{a}_2^\dagger(0)\hat{a}_2(0)\hat{a}_2^\dagger(0)\hat{a}_2^2(0) + \tilde{\omega}_1 k\hat{a}_1(0) \\ & \left. +2\tilde{g}\tilde{k}\hat{a}_1^\dagger(0)\hat{a}_1^2(0) + \tilde{k}^2 a_2(0)\right]\tau^2\end{aligned} \tag{5}$$

For convenience in numerical simulation, solutions (4)-(5) is expressed in terms of these dimensionless parameters defined as $\tilde{\omega}_1 = \omega_1/\omega_1$, $\tilde{\omega}_2 = \omega_2/\omega_1$, $\tilde{g} = g/\omega_1$ $\tilde{k} = k/\omega_1$ and $\tau = \omega_1 t$. These solutions only contain the term up to τ^2 only. The absence of higher-order terms may affect the sensitivity of lead to the loss of some essential physics.

2.2. Analytical Perturbative (AP) Method

In Heisenberg picture, the Heisenberg equation of motion has a general solution in the form of Baker-Campbell-Hausdorff (BCH) formula $\hat{a}_j = e^{i\hat{H}t}\hat{a}_j(0)e^{-i\hat{H}t}$. It is convenient to expand this solution using Maclaurin series as $\hat{a}_j = \hat{a}_j(0) + it[\hat{H}, \hat{a}_j(0)] - \frac{t^2}{2!}\left[\hat{H}, [\hat{H}, \hat{a}_j(0)]\right]$considering only the first three terms. The commutator terms in this expression can simply be obtained using the initial state of Eqs. (2)-(3). For this method, we employ the weak nonlinearity approximation, where the nonlinear coupling coefficients are restricted to have a linear power only, i.e., terms with $\tilde{g}^2$ and above are neglected. Following this mathematical process, the modes solutions take the following form.

$$\hat{a}_1 = A_1\hat{a}_1(0) + A_2\hat{a}_2(0) + A_3\hat{a}_1^\dagger(0)\hat{a}_1^2(0) + A_4\hat{a}_2^\dagger(0)\hat{a}_1^2(0) + A_5\hat{a}_1^\dagger(0)\hat{a}_1(0)\hat{a}_2(0) + A_6\hat{a}_2^\dagger(0)\hat{a}_2^2(0) \tag{6}$$

$$\hat{a}_2 = B_1\hat{a}_1(0) + B_2\hat{a}_2(0) + B_3\hat{a}_2^\dagger(0)\hat{a}_2^2(0) + B_4\hat{a}_1^\dagger(0)\hat{a}_2^2(0) + B_5\hat{a}_2^\dagger(0)\hat{a}_2(0)\hat{a}_1(0) + B_6\hat{a}_1^\dagger(0)\hat{a}_1^2(0) \tag{7}$$

However, these solutions are not complete yet, since the $A's$ and $B's$ coefficients remain unknown. These coefficients can be obtained directly from Maclaurin expansion. However, if we do so, The outcome of the AP method becomes similar to the STA method. Alternatively in the AP method, the A's and B's coefficients are assumed unknown functions of time with terms span from τ^0 to τ^∞. For this reason, the AP is said to be superior to the STA method. Therefore, solutions obtained using this method are more general compared with the STA method. Now, the A's and B's coefficients can be evaluated numerically by solving the following two sets of coupled differential equations;

$$dA_1/d\tau = i\left[\tilde{\omega}_1 A_1 + \tilde{k}B_1\right] \tag{8}$$

$$dA_2/d\tau = i\left[\tilde{\omega}_1 A_2 + \tilde{k}B_2\right] \tag{9}$$

$$dA_3/d\tau = i\left[\tilde{\omega}_1 A_3 + 2\tilde{g}\left|A_1\right|^2 A_1 + \tilde{k}B_6\right] \tag{10}$$

$$dA_4/d\tau = i\left[\tilde{\omega}_1 A_4 + 2\tilde{g}A_2^{\,*}A_1^{\,2}\right] \tag{11}$$

$$dA_5/d\tau = i\left[\tilde{\omega}_1 A_5 + 4\tilde{g}\left|A_1\right|^2 A_2\right] \tag{12}$$

$$dA_6/d\tau = i\left[\tilde{\omega}_1 A_6 + 2\tilde{g}\left|A_2\right|^2 A_2 + \tilde{k}B_3\right] \tag{13}$$

and

$$dB_1/d\tau = i\left[\tilde{\omega}_2 B_1 + \tilde{k}A_1\right] \tag{14}$$

$$dB_2/d\tau = i\left[\tilde{\omega}_2 B_2 + \tilde{k}A_2\right] \tag{15}$$

$$dB_3/d\tau = i\left[\tilde{\omega}_2 B_3 + 2\tilde{g}\left|B_2\right|^2 B_2 + \tilde{k}A_6\right] \tag{16}$$

$$dB_4/d\tau = i\left[\tilde{\omega}_2 B_4 + 2\tilde{g}B_1^{*}B_2^{2}\right] \tag{17}$$

$$dB_5/d\tau = i\left[\tilde{\omega}_2 B_5 + 4\tilde{g}\left|B_2\right|^2 B_1\right] \tag{18}$$

$$dB_6/d\tau = i\left[\tilde{\omega}_2 B_6 + 2\tilde{g}\left|B_1\right|^2 B_1 + \tilde{k}A_3\right] \tag{19}$$

These differential equations are obtained by substituting the assumed solutions (6)-(7) into the Heisenberg equation of motion (2)-(3) for the modes and collecting the identical terms from both sides. Here, we used the Fourth-order Runge Kutta method to solve these equations numerically.

2.3. *Squeezing in Kerr-Coupler*

Quantum mechanically, the electric field quadrature variances can be used to determine the level of noise in a light source. For squeezed light, the amount of noise must be lower than the quantum shot-noise inherited in the coherent states of light. The quadrature operator for the first and the second quadrature components are defined as $\hat{X}_j = \frac{1}{2}(\hat{a}_j + \hat{a}_j^{\dagger})$ and $\hat{Y}_j = -\frac{i}{2}(\hat{a}_j - \hat{a}_j^{\dagger})$, respectively. The variances for these quadrature operators can be derived as follow.

$$\left\langle (\Delta\hat{X}_j)^2\right\rangle, \left\langle (\Delta\hat{Y}_j)^2\right\rangle = \frac{1}{4}\left[1 + 2\left(\langle \hat{a}_j^{\dagger}\hat{a}_j\rangle - \langle \hat{a}_j\rangle\langle \hat{a}_j^{\dagger}\rangle\right) \pm \left(\langle \hat{a}_j^{2}\rangle - \langle \hat{a}_j\rangle^2 + \text{c.c.}\right)\right] \tag{20}$$

In this paper, we assume that the initial state of the field is coherent state. Hence, the initial expectation value of the mode-annihilation operators is defined as $\langle \hat{a}_j(0)\rangle = \alpha_j$, where α_j is the amplitude of the input coherent field. The mode solutions derived earlier from the STA method (4)-(5) and that of the AP method (6)-(7) will be used to evaluate the quadrature variances using Eq. (20). For the STA method, we substitute Eqs. (4)-(5) into (20) to get the following formulas,

$$\left\langle (\Delta\hat{X}_1)^2\right\rangle, \left\langle (\Delta\hat{Y}_1)^2\right\rangle = \frac{1}{4}\left[1 \pm 2\tilde{g}\tau\left(i\alpha_1^2 - 2\tilde{\omega}_1\alpha_1^2\tau - \tilde{k}\alpha_1\alpha_2\tau - \tilde{g}\alpha_1^2\tau - 8\tilde{g}\left|\alpha_1\right|^2\alpha_1^2\tau + \text{c.c.}\right)\right] \tag{21}$$

$$\left\langle (\Delta\hat{X}_2)^2\right\rangle, \left\langle (\Delta\hat{Y}_2)^2\right\rangle = \frac{1}{4}\left[1 \pm 2\tilde{g}\tau\left(i\alpha_2^2 - 2\tilde{\omega}_2\alpha_2^2\tau - \tilde{k}\alpha_2\alpha_1\tau - \tilde{g}\alpha_2^2\tau - 8\tilde{g}\left|\alpha_2\right|^2\alpha_2^2\tau + \text{c.c.}\right)\right] \tag{22}$$

where c.c. is complex conjugate. For the AP method, the quadrature variance is derived by substituting Eqs. (6)-(7) into (20), and we obtain

$$\left\langle (\Delta\hat{X}_1)^2\right\rangle, \left\langle (\Delta\hat{Y}_1)^2\right\rangle = \frac{1}{4}\left[1 \pm \left[(A_3A_1 + A_4A_2)\alpha_1^2 + A_5A_1\alpha_1\alpha_2 + A_6A_2\alpha_2^2 + \text{c.c.}\right]\right] \tag{23}$$

$$\left\langle (\Delta\hat{X}_2)^2\right\rangle, \left\langle (\Delta\hat{Y}_2)^2\right\rangle = \frac{1}{4}\left[1 \pm \left[(B_4B_1 + B_3B_2)\alpha_2^2 + B_5B_2\alpha_2\alpha_1 + B_6B_1\alpha_1^2 + \text{c.c.}\right]\right]. \tag{24}$$

Using these equations (21)-(24), the condition of squeezing is given by $\langle (\Delta\hat{X}_j)^2\rangle < \frac{1}{4}$ or $\langle (\Delta\hat{Y}_j)^2\rangle < \frac{1}{4}$. In deriving Eqs. (21)-(24), the following identities of quantum mechanics are essential to convert the expression into the normal-ordered operator configuration: $\hat{a}_j(\hat{a}_j^\dagger)^m = m(\hat{a}_j^\dagger)^{m-1} + (\hat{a}_j^\dagger)^m \hat{a}_j$ and $\hat{a}_j^m \hat{a}_j^\dagger = m\hat{a}_j^{m-1} + \hat{a}_j^\dagger \hat{a}_j^m$, where m is an integer number. Finally, these equations are solved numerically and the graphical results are presented in the next section.

3. Result and Discussion

Throughout the investigation, some of the parameters are set to be constant. These parameters are $\alpha_1 = 1$ $\alpha_2 = 1$ $\tilde{\omega}_1 = 1$ and $\tilde{g} = 0.01$. For all the graphs presented here, squeezing is identified whenever the field oscillates below the shot noise. For a clear indication, the shot noise is illustrated by the horizontal 'black solid line'. Figure 2a shows the quadrature fluctuation obtained using the analytical perturbative method. As seen, the squeezing occurs in a periodical manner, hence it is not continuous as indicated by the STA method. Plus, if squeezing exists in the first quadrature component, it will not exists in the other quadrature component at the same time, and vice versa. This is due to the Heisenberg uncertainty relation. Overall, squeezing is possible in both quadrature components. In Fig. 2b, we notice the quadratic-like behavior of quadrature variances resulted from the STA formulation without oscillation is detected. Because of this limitation, STA fails to detect the existence of squeezing in the second quadrature component. This confirms our observation that the STA method may ignore some important physics. Figure 2c shows the combination of the two previous graphs. Here, we can see a significant difference between the AP method and the STA method at long evolution distances. As time increases, the disagreement between these two results becomes more obvious. Nevertheless, the two methods highly agree with each other at the very beginning. Therefore, we can say that the accuracy of the STA method is limited for a very short time only. the failure of the STA method at a longer evolution time may be due to the elimination of terms beyond the second-order terms with τ^2.

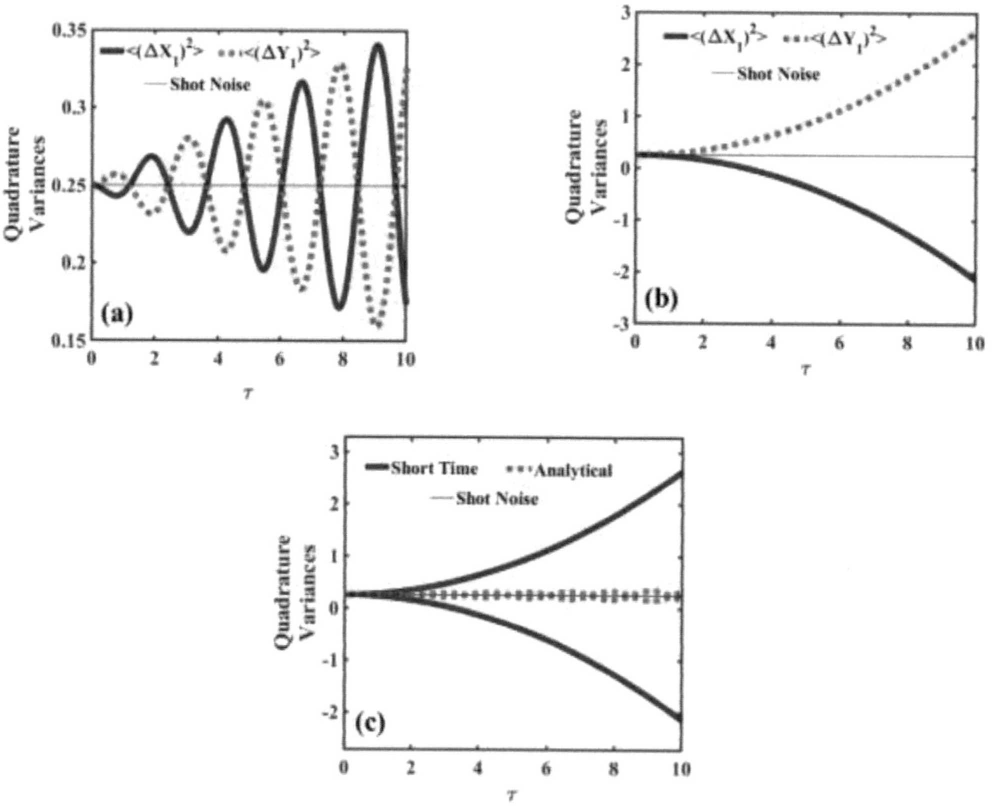

Figure 2. Time evolution of the quadrature variances for the first waveguide-mode using (a) the AP method and (b) the STA method. A combination of these two graphs is shown in (c). ΔX_1^2 and ΔY_1^2 are

first and the second quadrature component of the field mode respectively. The optical mode in the second waveguide shows exact similar results. The input parameters used $\Delta\tilde{\omega} = 0$ and $\tilde{k} = 0.3$.

In Fig. 3, squeezing for different values of linear coupling coefficient $\tilde{k}$ is shown. Here, we only show the first quadrature component ΔX_1^2 of the first waveguide mode. In the AP method, as $\tilde{k}$ increases, the squeezing occurs more frequently (see Fig. 3a). This is true, since the quadrature variances exhibit dense oscillations at higher values $\tilde{k}$ due to the rapid exchange of energy between the waveguides. For the considered time range, the maximal squeezing appears four times and six times at $\tilde{k} = 0.1$ and $\tilde{k} = 0.8$, respectively. Although the frequency of the quadrature oscillation increases, the amplitude of the graph remains unchanged. This indicates that the strength of the squeezed light is not affected by the value of $\tilde{k}$. As mention earlier, energy is conserved in the interaction process, since the energy gain by one channel is balanced with the energy loss in the other channel. Therefore, by strengthening the linear coupling interaction, no extra energy will be added to the system. On the other side, In the STA method (Fig. 3b), the curve of $\tilde{k} = 0.8$ lies lower than the curve of $\tilde{k} = 0.1$. This indicates that a strong linear coupling yields a stronger squeezing compared with weak linear coupling. This outcome opposes the AP results. Furthermore, the second quadrature component ΔY_1^2 oscillates completely above the shot-noise level as observed previously in Fig. 2.

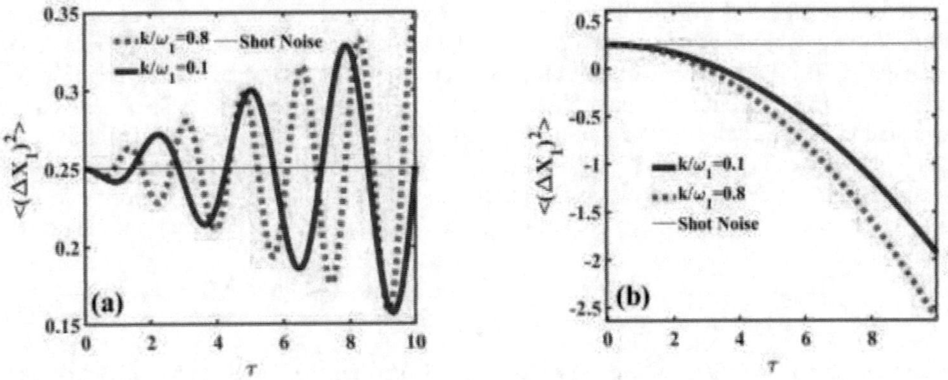

Figure 3. Temporal evolution of the first quadrature variance in the first waveguide mode using (a) the AP method and (b) the STA method for different values of linear coupling coefficient. Other parameters remain as in Fig. 2.

When we introduce a frequency mismatch between both modes, the pattern of the graph generated by the AP method changes (Fig. 4a). In this case, the modes propagate at different phases. Therefore, the interaction between the modes takes place differently compared with the case of frequency matching. However, this change in quadrature pattern is not predicted by the STA method (see Fig. 4b). As we can see here, both graphs are overlapped with each other to indicate that the interaction occurs similarly for both conditions. Therefore, the STA method seems not suitable to examine the squeezed light under the presence of frequency mismatch.

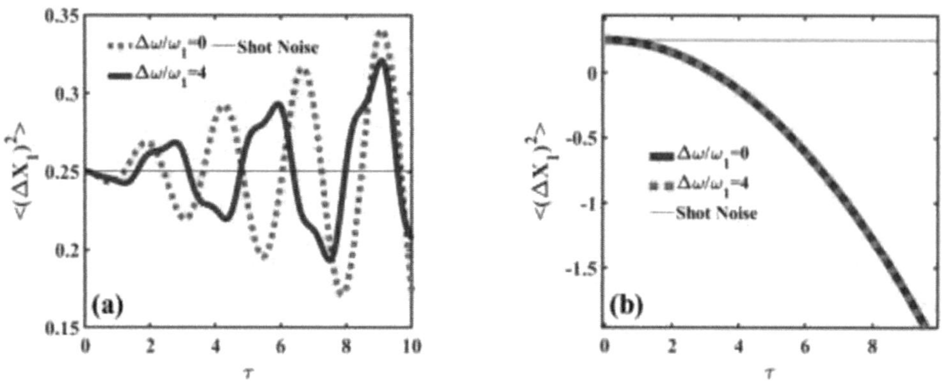

Figure 4. Squeezing generated by (a) AP method and (b) STA method for a different frequency mismatch in the first mode. The linear coupling coefficient is set as $\tilde{k} = 0.3$.

4. Conclusion

In conclusion, the AP method and the STA method have been used to study the detection of single-mode squeezed states of light generated in Kerr nonlinear coupler composed of two waveguides. Special focus was given to the comparison between these two mathematical methods. Both methods describe the system in Heisenberg picture in quantum mechanics, where the dynamics are characterized by Heisenberg equation of motion. In the STA method, the Heisenberg equation of motion is solved utilizing the short-time approximation, while the AP method assumes a weak nonlinearity. Mathematically, we have seen that the solution of the AP method remains more general compared with the STA method. The STA method does not show oscillations in both quadrature variances of the propagating modes which contradict the oscillatory nature of light. Besides, the existence of squeezing in the second quadrature component cannot be detected using the STA method. Nevertheless, the STA method is highly agreeable with the AP method at a very early evolution. The STA method predicts the amplification of the squeezed signal when the linear coupling is strengthened as opposing to the AP method. Furthermore, The STA method also fails to detect the effect of frequency mismatch on the propagating modes.

References

[1] M. Mehmet and H. Vahlbruch, "The Squeezed Light Source for the Advanced Virgo Detector in the Observation Run O3," p. 10, 2020.

[2] K. Liu, C. Cai, J. Li, L. Ma, H. Sun, and J. Gao, "Squeezing-enhanced rotating-angle measurement beyond the quantum limit," *Appl. Phys. Lett.*, p. 6, 2018.

[3] A. Furusawa, "Unconditional Quantum Teleportation," *Science*, vol. 282, no. 5389, pp. 706–709, Oct. 1998, DOI: 10.1126/science.282.5389.706.

[4] S. L. Braunstein and H. J. Kimble, "Teleportation of Continuous Quantum Variables," *Physical Review Letters*, vol. 80, no. 4, p. 4, 1998.

[5] M. A. Bashar, M. A. Chowdhury, R. Islam, M. S. Rahman, and S. K. Das, "A Review and Prospects of Quantum Teleportation," in *2009 International Conference on Computer and Automation Engineering*, Bangkok, Mar. 2009, pp. 213–217. DOI: 10.1109/ICCAE.2009.77.

[6] J. Shapiro, H. Yuen, and A. Mata, "Optical communication with two-photon coherent states--Part II: Photoemissive detection and structured receiver performance," *IEEE Trans. Inform. Theory*, vol. 25, no. 2, pp. 179–192, Mar. 1979, DOI: 10.1109/TIT.1979.1056033.

[7] H. P. Yuen and J. H. Shapiro, "Generation and detection of two-photon coherent states in degenerate four-wave mixing," *Opt. Lett.*, vol. 4, no. 10, p. 334, Oct. 1979, DOI: 10.1364/OL.4.000334.

[8] S. L. Braunstein and H. J. Kimble, "Dense coding for continuous variables," *PHYSICAL REVIEW A*, p. 4.

[9] C. M. Caves, "Quantum-mechanical noise in an interferometer," *Phys. Rev. D*, vol. 23, no. 8, pp. 1693–1708, Apr. 1981, DOI: 10.1103/PhysRevD.23.1693.

[10] The LIGO Scientific Collaboration, "A gravitational wave observatory operating beyond the quantum shot-noise limit," *Nature Phys*, vol. 7, no. 12, pp. 962–965, Dec. 2011, DOI: 10.1038/nphys2083.

[11] N. Treps *et al.*, "Surpassing the Standard Quantum Limit for Optical Imaging Using Nonclassical Multimode Light," *Phys. Rev. Lett.*, vol. 88, no. 20, p. 203601, May 2002, DOI: 10.1103/PhysRevLett.88.203601.

[12] M. I. Kolobov and C. Fabre, "Quantum Limits on Optical Resolution," *Phys. Rev. Lett.*, vol. 85, no. 18, pp. 3789–3792, Oct. 2000, DOI: 10.1103/PhysRevLett.85.3789.

[13] M. A. Taylor *et al.*, "Biological measurement beyond the quantum limit," *Nature Photon*, vol. 7, no. 3, pp. 229–233, Mar. 2013, DOI: 10.1038/nphoton.2012.346.

[14] M. Lassen *et al.*, "Quantum optical coherence can survive photon losses using a continuous-variable quantum erasure-correcting code," *Nature Photon*, vol. 4, no. 10, pp. 700–705, Oct. 2010, DOI: 10.1038/nphoton.2010.168.

[15] T. Aoki *et al.*, "Quantum error correction beyond qubits," *Nature Phys*, vol. 5, no. 8, pp. 541–546, Aug. 2009, DOI: 10.1038/nphys1309.

[16] M. S. M. Hanapi, A.-B. M. A. Ibrahim, R. Julius, and H. Eleuch, "The Quantum Kerr Nonlinear Coupler: The Analytical Versus Phase-Space Method," *Canadian Journal of Physics*, p. 17.

[17] R. Julius, A.-B. M. A. Ibrahim, H. Eleuch, and P. K. Choudhury, "Sub-Poissonian photon squeezing and entanglement in optical chain second harmonic generation," *Journal of Modern Optics*, vol. 66, no. 10, pp. 1129–1138, Jun. 2019, DOI: 10.1080/09500340.2019.1603408.

[18] R. Julius, M. Zahirzai, A.-B. M. A. Ibrahim, H. Eleuch, and P. K. Choudhury, "Single- and compound-mode squeezing in nonlinear coupler with frequency mismatch," *Journal of Electromagnetic Waves and Applications*, pp. 1–15, Dec. 2019, DOI: 10.1080/09205071.2019.1702591.

[19] R. Julius, A.-B. M. A. Ibrahim, P. K. Choudhury, and H. Eleuch, "Quantum states generation in multichannel directional coupler with second-order nonlinearity," *Optik*, vol. 186, pp. 212–220, Jun. 2019, DOI: 10.1016/j.ijleo.2019.03.027.

[20] R. Julius, A.-B. M. A. Ibrahim, P. K. Choudhury, and H. Eleuch, "On the nonclassical dynamics of cavity-assisted four-channel nonlinear coupler," *Chinese Phys. B*, vol. 27, no. 11, p. 114206, Nov. 2018, DOI: 10.1088/1674-1056/27/11/114206.

[21] M. S. M. Hanapi, A.-B. M. A. Ibrahim, and P. K. Choudhury, "On the perturbative approach to analyse entanglement in two-channel Kerr nonlinear coupler," *Optik*, vol. 243, p. 167420, Oct. 2021, DOI: 10.1016/j.ijleo.2021.167420.

[22] B. A. Umarov, A.-B. M. A. Ibrahim, and M. R. B. Wahiddin, "Quantum Dynamics in Kerr Nonlinear Coupler Via Phase-Space Representation," in *Nonlinearity and Disorder: Theory and Applications*, F. Abdullaev, O. Bang, and M. P. Sørensen, Eds. Dordrecht: Springer Netherlands, 2001, pp. 341–349. DOI: 10.1007/978-94-010-0542-5_29.

[23] R. Julius, A.-B. M. A. Ibrahim, M. K. Abd-Rahman, and P. K. Choudhury, "Quantum dynamics of a four-channel Kerr nonlinear directional coupler," *Opt Rev*, vol. 25, no. 5, pp. 563–570, Oct. 2018, DOI: 10.1007/s10043-018-0448-x.

[24] R. Julius, A.-B. M. A. Ibrahim, and M. S. M. Deni, "Quantum state generation in a four-mode Kerr nonlinear directional coupler," *Laser Phys.*, vol. 24, no. 3, p. 035202, Mar. 2014, DOI: 10.1088/1054-660X/24/3/035202.

[25] A.-B. M. A. Ibrahim, B. A. Umarov, and M. R. B. Wahiddin, "Squeezing in the Kerr nonlinear coupler via phase-space representation," *Phys. Rev. A*, vol. 61, no. 4, p. 043804, Mar. 2000, DOI: 10.1103/PhysRevA.61.043804.

[26] A. Politi, J. C. F. Matthews, and J. L. O'Brien, "Shor's Quantum Factoring Algorithm on a Photonic Chip," *Science*, vol. 325, no. 5945, pp. 1221–1221, Sep. 2009, DOI: 10.1126/science.1173731.

[27] P. Mandal and S. Midda, "All optical method of developing OR and NAND logic system based on nonlinear optical fiber couplers," *Optik*, vol. 122, no. 20, pp. 1795–1798, Oct. 2011, DOI: 10.1016/j.ijleo.2010.10.041.

[28] H. W. Li *et al.*, "Reconfigurable controlled two-qubit operation on a quantum photonic chip," p. 9.

[29] J. C. F. Matthews, A. Politi, A. Stefanov, and J. L. O'Brien, "Manipulation of multiphoton entanglement in waveguide quantum circuits," *Nature Photon*, vol. 3, no. 6, pp. 346–350, Jun. 2009, DOI: 10.1038/nphoton.2009.93.

[30] K. Mukherjee and P. C. Jana, "Squeezing and Entanglement in Quadratically-Coupled Optomechanical System," p. 13.

[31] K. Mukherjee and P. C. Jana, "Higher-Order Squeezing of a Generic Quadratically-Coupled Optomechanical System," *IOSR Journal of Applied Physics*, vol. 9, no. 01, pp. 90–95, Feb. 2017, DOI: 10.9790/4861-0901019095.

[32] A. Mukhopadhyay, B. Sen, K. Thapliyal, S. Mandal, and A. Pathak, "Interaction of light and semiconductor can generate quantum states required for solid-state quantum computing: entangled, steered and other nonclassical states," *Quantum Inf Process*, vol. 18, no. 7, p. 234, Jul. 2019, DOI: 10.1007/s11128-019-2344-0.

[33] B. Sen, V. Peřinová, A. Lukš, J. Peřina, and J. Křepelka, "Sub-shot noise photon-number correlation in stimulated and spontaneous Raman processes," *J. Phys. B: At. Mol. Opt. Phys.*, vol. 44, no. 10, p. 105503, May 2011, DOI: 10.1088/0953-4075/44/10/105503.
[34] B. Sen and S. Mandal *, "Squeezed states in spontaneous Raman and in stimulated Raman processes," *Journal of Modern Optics*, vol. 52, no. 13, pp. 1789–1807, Sep. 2005, DOI: 10.1080/09500340500072984.
[35] K. Thapliyal, A. Pathak, B. Sen, and J. Peřina, "Lower- and higher-order nonclassical features in non-degenerate hyper-Raman processes," *Optics Communications*, vol. 444, pp. 111–119, Aug. 2019, DOI: 10.1016/j.optcom.2019.03.044.
[36] S. K. Giri, B. Sen, A. Pathak, and P. C. Jana, "Higher order two-mode and multi-mode entanglement in Raman processes," *Phys. Rev. A*, vol. 93, no. 1, p. 012340, Jan. 2016, DOI: 10.1103/PhysRevA.93.012340.
[37] M. Das, B. Sen, A. Ray, and A. Pathak, "Lower order and higher order entanglement in $87Rb$ $5S-5P-5D$ hyperfine manifold modeled as a four-wave mixing process," *ANNALEN DER PHYSIK*, vol. 530, no. 1, p. 1700160, Jan. 2018, DOI: 10.1002/andp.201700160.
[38] S. K. Giri, K. Thapliyal, B. Sen, and A. Pathak, "Nonclassicality in an atom–molecule Bose–Einstein condensate: Higher-order squeezing, antibunching and entanglement," *Physica A: Statistical Mechanics and its Applications*, vol. 466, pp. 140–152, Jan. 2017, DOI: 10.1016/j.physa.2016.09.004.
[39] G. S. Kumar, S. Biswajit, and J. P. Chandra, "General Higher Order Intermodal Antibunching in two-mode Bose Einstein Condensates," vol. 5, p. 4, 2016.
[40] N. Alam, K. Thapliyal, A. Pathak, B. Sen, A. Verma, and S. Mandal, "Bose-condensed optomechanical-like system and a Fabry–Perot cavity with one movable mirror: quantum correlations from the perspectives of quantum optics," *Eur. Phys. J. D*, vol. 73, no. 7, p. 139, Jul. 2019, DOI: 10.1140/epjd/e2019-90448-x.

ACKNOWLEDGEMENT

A.M.A.I. acknowledges the financial support by RMC, UiTM, Shah Alam, Selangor, Malaysia. Grant ID: 600-RMC/MYRA 5/3/LESTARI (069/2020).

COPPER-BASED 3D PRINTER FILAMENT AS PASSIVE Q-SWITCHER

N. N. A. A. N Zain [1], N. A. H. Jasni [1], W. M. F. W Nawawi [2] and F. Ahmad [1*]

[1] Malaysia- Japan International Institute of Technology, Universiti Teknologi Malaysia, Jalan Sultan Yahya Petra, 54100 Kuala Lumpur, Malaysia
[2] Department of Biotechnology Engineering, International Islamic University Malaysia, Kuala Lumpur, Malaysia

*Corresponding author's email: fauzan.kl@utm.my

Abstract. Using a copper-based saturable absorber (SA) and chitin as a biocompatible host polymer, this research effectively demonstrated the production of passive Q-switched in an erbium-doped fibre laser (EDFL). A Q-switched lasing was generated with a 100 kHz repetition rate and a pulse width of 4.60 μs. High pulse energy of 17.19 nJ was generated, with a respective instantaneous peak power of 3.51 mW.

1. Introduction

Researchers have recently become interested in metal nanoparticles, such as copper, for the next generation of SAs for use in passive Q-switching [1]. This is due to the Localized Surface Plasmon Resonance (LSPR) feature of metal nanomaterials, which is caused by the collective oscillation of conduction electrons in metal nanoparticles [8]. Copper has a broad saturable absorption range and a fast response time [2, 3]. Copper-based pulse laser generation are reported to operate at broad operating wavelength [9-12] with various source materials such as copper pellets [9], copper nanoparticles [10], copper oxide nano powder [11], and copper nanowires [12]. Using copper-based 3D printer filament as the starting material and Chitin biocompatible host polymer, this work provided an alternate method for developing copper-based passive Q-switchers. This method aims to investigate the use of metal-based 3D printer filament in optics. Because of the compatibility of the solvent used to dissolve the metal base, the chitin biocompatible polymer was chosen.

2. Methodology

2.1. Copper-chitin SA fabrication

3D Gadgets provided the copper-based 3D printer filament (Magma). The diameter of the 3D printer filament is 1.75mm +/- 0.03mm, the length is 155mm, the heated bed temperature is 0- 60°C, the printing temperature is 190°C - 210°C, the net weight is 0.50 kg, and the material is Polylactic Acid (PLA) with 15% copper powder. The size of 3D printer filament was red for easier construction. . The size of 3D printer filament was reduced from 1.75 mm to 400 m by extrusion at 210°C through a 0.4 mm extruder nozzle for easier fabrication process. To dissolve the copper-PLA 3D printer filament, 25 mg of reduced diameter copper filament was thoroughly combined with 1 ml tetrahydrofuran (THF) solvent, to produced copper-PLA-THF suspension. Wan Nawawi et al. [4] described the production of Chitin as a host polymer using oyster mushrooms as the source of chitin. To develop a copper-chitin-based passive

2075 (2021) 012003 doi:10.1088/1742-6596/2075/1/012003

saturable absorber, the copper-PLA-THF suspension was completely mixed with chitin suspension in a one-to-one ratio at room temperature trough ultrasonication. Around 5 mL of the copper-PLA-THF-Chitin homogeneous suspension was poured into a dish and dry for 36 hours at ambient, where the PLA-THF evaporated, leaving a copper-chitin-based film. The free-standing copper-chitin based film will be peeled off and divided into 1 mm x 1 mm squares before being integrated in ring laser cavity for pulsed laser generation.

2.2. Experimental setup

As illustrated in Figure 1, a 980 nm laser diode (LD) was utilised to pump the erbium-doped fibre laser (EDFL) ring cavity. A 980/1550 nm wavelength division multiplexer (WDM) pushed the laser towards the gain medium, which was a 1.7 m erbium-doped fibre gain medium. An isolator was installed in the cavity to ensure unidirectional light propagation. Before being integrated into the cavity as a passive Q-switcher, the free-standing copper chitin was sandwiched between two fibre ferrules with the help of index matching gel. For the output observation, the light passes via a 95/5 output coupler, with 95% of the signal remaining in the cavity and the remaining 5% tapped out as data gathering output. A Yokogawa AQ6370B Optical Spectrum Analyzer (OSA), an optical power meter (OPM), a Tektronix MDO3024 Mixed Domain Oscilloscope with Thorlabs DET01CFC 460 kHz bandwidth photodetector (PD) were used to monitor the signal.

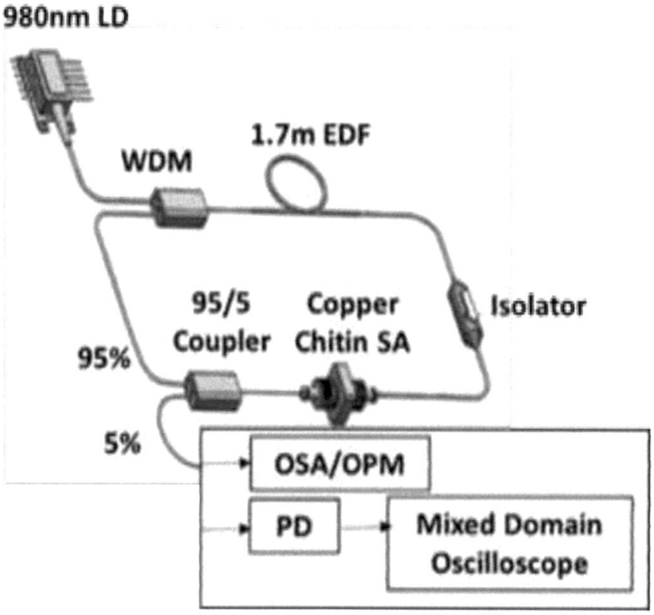

Figure 1 Integration of copper-chitin SA in ring laser cavity

3. Result and discussion

The Q-switching wavelengh of copper chitin SA in the laser cavity is 1559.316 nm, as illustrated in Figure 2. The threshold power of the Q-switched is at 38.7 mW and up maximum the pump power of 71.8 mW. Because the SAs have been saturated, Q-switched pulses cannot be created above 71.8 mW pump power, and only limited signal absorption can be achieved at higher pump power [5].

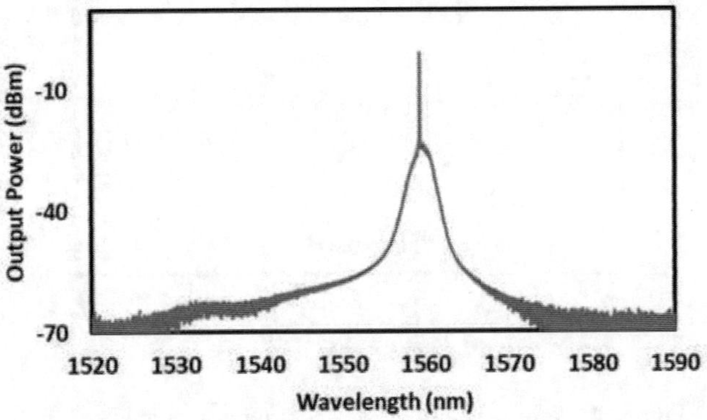

Figure 2 The optical spectrum at 71.8 mW of pump power.

A continuous Q-switching operation with no amplitude modulation was shown in Figure 3.Figure 3(a) depicts the Q-switching operation's pulse train repetition rates. At an input pump power of 71.8 mW, the copper chitin SA produced a pulse train with a 100 kHz repetition rate and a 10 μs pulse separation. Figure 3(b) illustrates a pulse width of 4.60 μs at the 71.8 mW.

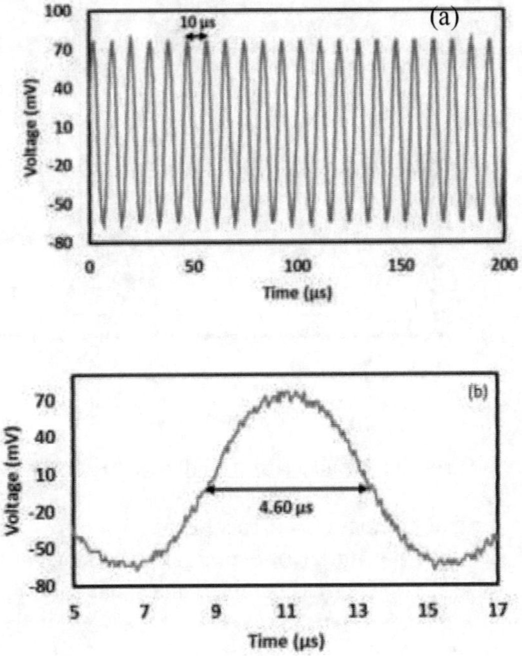

Figure 3 A continuous Q-switching operation (a) repetition rate (b) pulse width at 71.8mW.

Figure 4 illustrates that when the pump power increases, the repetition rate climbs almost linearly until it approaches 100 kHz, which is typical of Q-switched operations [6]. The pulse width decreased within input power from 38.7 mW to 71.8 mW, decreasing from 7.57 μs to 4.60 μs. Increasing pump power should theoretically provide more gain in producing SA saturation, resulting in a higher repetition rate and smaller pulse width [6][7].

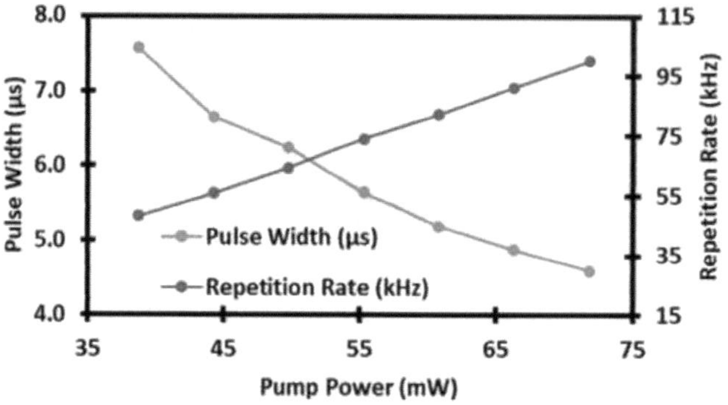

Figure 4 Pump power against repetition rate and pulse width.

Figure 5 illustrates how the instantaneous pulse energy and peak power were calculated using the observed repetition rate and pulse width [13]. Pumping power was raised from 38.7 to 71.8 mW, resulting in enhanced pulse energy and peak power.

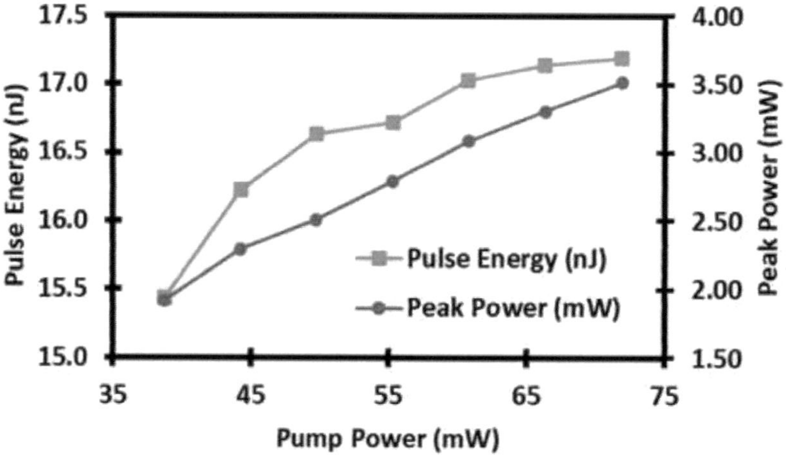

Figure 5 Instantaneous pulse energy and peak power versus input pump power.

Figure 6 shows how a mixed domain oscilloscope was used to verify the stability of the generated Q-switched pulse. At 71.8 mW, the initial beat note at 98 kHz was 80.18 dB, indicating good pulse stability. The signal-to-noise ratio (SNR) was higher than that of copper nanoparticles, which was 54 dB [3].

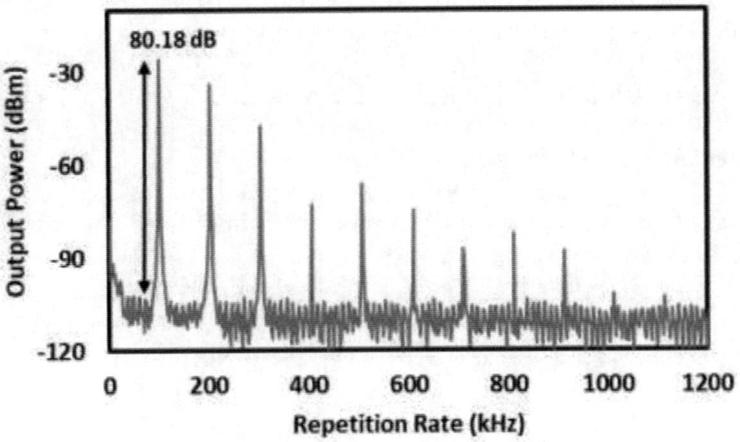

Figure 6 Signal-to-noise ratio

4. Conclusion

Copper-chitin SA was used to demonstrate a passively Q-switched EDFL with a highest repetition rate of 100 kHz, the shortest pulse width of 4.60 s, and a pulse energy of 17.19 nJ at input pump power of 71.8 mW. The pulse's stability can also be seen in the mixed domain oscilloscope with signal-to-noise ratio of around 80.18 dB. The copper-based 3D printer filament in chitin biocompatible host polymer revealed the ability to operate as a passive Q-switcher in this study.

References

[1] Ismail E, Ahmad F, Ambran S, Latiff A, Harun S. Q-switched erbium-doped fiber lasers based on copper nanoparticles saturable absorber. Journal of Physics: Conference Series. 2019;1371:012028.

[2] Lu S, Miao L, Guo Z, Qi X, Zhao C, Zhang H et al. Broadband nonlinear optical response in multi-layer black phosphorus: an emerging infrared and mid-infrared optical material. Optics Express. 2015;23(9):11183.

[3] Guo H, Feng M, Song F, Li H, Ren A, Wei X et al. Q-Switched Erbium-Doped Fiber Laser Based on Silver Nanoparticles as a Saturable Absorber. IEEE Photonics Technology Letters. 2016;28(2):135-138.

[4] Fazli Wan Nawawi W, Lee K, Kontturi E, Murphy R, Bismarck A. Chitin Nanopaper from Mushroom Extract: Natural Composite of Nanofibers and Glucan from a Single Biobased Source. ACS Sustainable Chemistry & Engineering. 2019;7(7):6492-6496.

[5] Kiat YY. Chemical Synthesis And Characterization of Graphene Oxide For Use As Saturable Absorber and Broadband Polarizer [dissertation].Kuala Lumpur, Malaysia : University of Malaya; 2015.

[6] Holzwarth A. Applications of ultrafast laser spectroscopy for the study of biological systems. Quarterly Reviews of Biophysics. 1989;22(3):239-326.

[7] Degnan J. Optimization of passively Q-switched lasers. IEEE Journal of Quantum Electronics. 1995;31(11):1890-1901.

[8] Murray, W. A., & Barnes, W. L. (2007). Plasmonic materials. Advanced materials, 19(22), 3771-3782.

[9] Muhammad, A.R, Ahmad, M., Zakaria, R., Rahim, H., Yusoff, S., Hamdan, K., Yusof, H., Arof, H., & Harun, S.W. (2017). Q-Switching Pulse Operation in 1.5-μm Region Using Copper Nanoparticles as Saturable Absorber. Chinese Physics Letters, 34(3), 034205.

[10] Muhammad, A.R, Ahmad, M., Zakaria, R., Yupapin, P., Harun, S.W., & Yasin, M. (2019). Mode-locked thulium doped fibre laser with copper thin film saturable absorber. Journal of Modern Optics, 66(13), 1381-1385.

[11] Sadeq, S. A., Al-Hayali, S. K., Harun, S. W., & Al-Janabi, A. (2018). Copper oxide nanomaterial saturable absorber as a new passive Q-switcher in erbium-doped fiber laser ring cavity configuration. Results in Physics, 10, 264-269.

[12] Ismail E, Ahmad F, Shafie S, Yahaya H, Latif A, Muhammad F. Copper nanowires based mode-locker for soliton nanosecond pulse generation in erbium-doped fiber laser. Results in Physics. 2020;18:103228.

[13] Zain N N A A N, Zuikafly S N F, Jasni N A H, Nawawi W M F W and Ahmad F. Passively Q-Switched Tri-Wavelength Erbium-Doped Fiber laser with Aluminium-Based Saturable Absorber. Results in IEEE International Conference on Semiconductor Electronics (ICSE). 2020:128-131

Silver Nanoparticles Embedded in Polyvinyl Alcohol Based Passive Saturable Absorber

N N M Rashid[1], M Q Lokman[1], N A Nordin[1], H Yahaya[1] and F Ahmad[1*]

[1] Malaysia-Japan International Institute of Technology, Universiti Teknologi Malaysia, Jalan Sultan Yahya Petra, 54100, Kuala Lumpur, Malaysia

*Corresponding author's email: fauzan.kl@utm.my

Abstract. Silver nanoparticles (AgNP), a novel candidate of noble metal nanoparticles, which has great potential for pulsed laser applications due to strong saturation absorption behaviour. This study demonstrated the passively Q-switched laser operating within an Erbium-doped fibre laser cavity using AgNP saturable absorber (SA). The AgNP was synthesized via the chemical reduction method and the average size of nanoparticles was around 52.39 nm. The AgNP was incorporated with polyvinyl alcohol (PVA) polymer to form an AgNP-PVA composite SA. The proposed experimental works generated single pulse energy and peak power of about 66.14 nJ and 18.71 mW corresponds to the repetition rate and pulse width of 62.89 kHz and 3.32 μs.

1. Introduction

The development of saturable absorber (SA) technology has been practically parallel with the evolution of fibre lasers. 2D family, including graphene, MXenes, transition metal dichalcogenides, and topological material has expanded significantly in recent years. In 2D materials, the bandgap is an important characteristic for SA selection to identify the saturable absorption at the responding operation wavelength. Supposing, for the 2D family, the saturable absorption process occurs when the absorption of light reduces with the increasing light intensity just above the respective SA bandgap [1]. The bandgap of the SA determines the operation wavelength and the larger bandgap induced the weak electron-hole interaction due to the long separation of conduction and valence band [1]. Nonetheless, in noble metal nanoparticles, especially silver nanoparticles (AgNP), the saturable absorption can happen in two possibilities: the ground state plasmon bleaching due to intrinsic electron dynamics and intensity-dependent shift of the plasmon resonance induces the strong saturation behaviour [2,3].

In this work, we reported the Q-switched pulsed fibre laser using AgNP-PVA composite thin film as SA at Erbium operating wavelength. The AgNP was synthesized using the chemical reduction method and incorporated into PVA polymer composite. The obtained results show that the pulsed able to generate the pulse energy, peak power, repetition rate, and pulse width of 66.14 nJ, 18.71 mW, 62.89 kHz, and 3.32 μs, respectively.

2. Methodology

All the chemicals were used without any further purifications. The AgNP was synthesized using chemical reduction process in which sodium nitrate was used as the reducing agent. The average AgNP sizes are 52.39 nm with d-spacing of 2.17 Å (200 plane) shows that sodium citrate was successfully reduced from Ag salts to Ag^+ ions and resulting in the formation of Ag cluster. Additional information on characterization and methodology of AgNP, one can refer from our previous report [4,5]. Briefly, the

AgNP-PVA SA was prepared by separating the AgNP and excess sodium citrate using centrifuges. The process was repeated at least three times in order to produced high purity AgNP. Then, a known amount of AgNP was added into 5 mL PVA solution (10 mg/mL: PVA/DI water) and ultrasonicated for an hour followed by magnetic stirring for 3 h. Finally, the AgNP-PVA was transferred to the surface of a clean plastic mould and dried under ambient for 72 hours.

To generate the Q-switched pulsed, AgNP-PVA SA was further cut into small pieces of 1 mm^2 and transferred onto fibre ferrules to function as SA. The arrangement of the Q-switched pulsed laser used in this experiment was identical to our previous work reported elsewhere [5]. Worth mentioning that, our previous report focused on the Q-switched using Erbium-doped fibre (gain medium) with a length of 2.4 m. However, in this experiment, we used a gain medium with a length of 1.7 m to observed the changes of Q-switched pulsed laser performance. Figure 1 (a) and (b) shows the physical image of AgNP-PVA SA and Ag-PVA onto the fibre ferrule end.

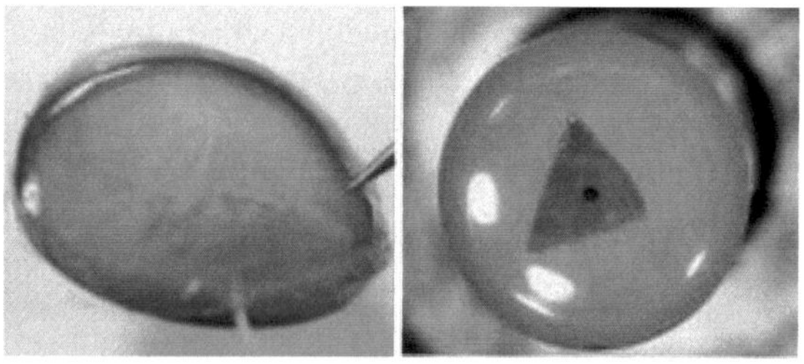

(a) (b)

Figure 1. Physical image of AgNP-PVA SA after 3 days drying under room temperature and (b) transferred onto a FC/PC fibre end.

3. Results and Discussions

The stable Q-switched operation was obtained at the pump power of 19.4 mW. However, the pulsed was observed to be distorted and vanished when the pump power increased exceeded 121.5 mW due to the oversaturation of the SA at high input intensity. In addition, pure PVA composites was tested in the Q-switched laser cavity and no pulse was observed shows that AgNP was operated in a Q-switched mode. Figure 2 (a) shows the Q-switched pulse train at a maximum pump power of 121.5 mW with pulse-pulse separation of 15.84 μs retrieved from the digital oscilloscope. Figure 2 (b) depicts the single pulse profile of the pulse train at a maximum pump power of 121.5 mW. The single pulse shows a symmetrical Gaussian-like shape with full-width at half maximum of 3.32 μs which is a typical time regime of Q-switched pulsed laser operation.

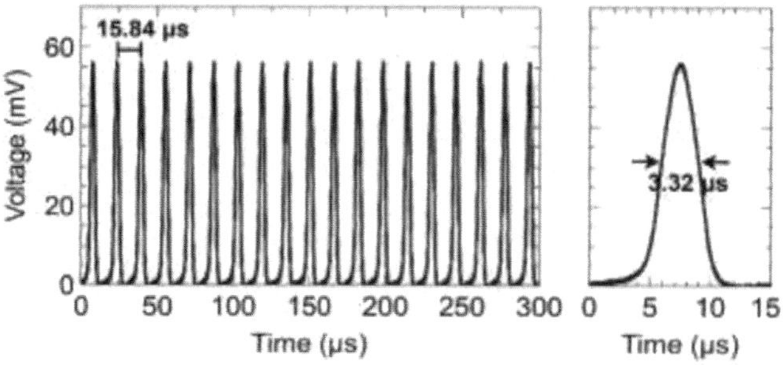

(a) (b)

Figure 2. (a) Typical pulse train and pulse width at pump power of 121.5 mW and (b) the pulse-to-pulse separation in 15.84 μs with single pulse width of 3.32 μs.

The repetition rate, pulse width, pulse energy, and peak power as a function of pump power of the Q-switched pulsed are presented in figure 3 (a) and (b). The results show that as the pump power increased, the repetition rate was monotonically increased from 24.75 to 62.89 kHz and the pulse widths were decreased as the pump power increases show the typical Q-switched operation. This is due to the higher pump power allowed more gain to saturate the AgNP-PVA SA in a shorter period of time, which leads to faster bleaching of AgNP-PVA. As a result, the repetition rate has increased. By tuning the pump power, the pulse energy and peak power were observed to increased. At the maximum pump power of 121.5 mW, the calculated single pulse energy is 66.14 nJ with instantaneous peak power of 18.71 mW.

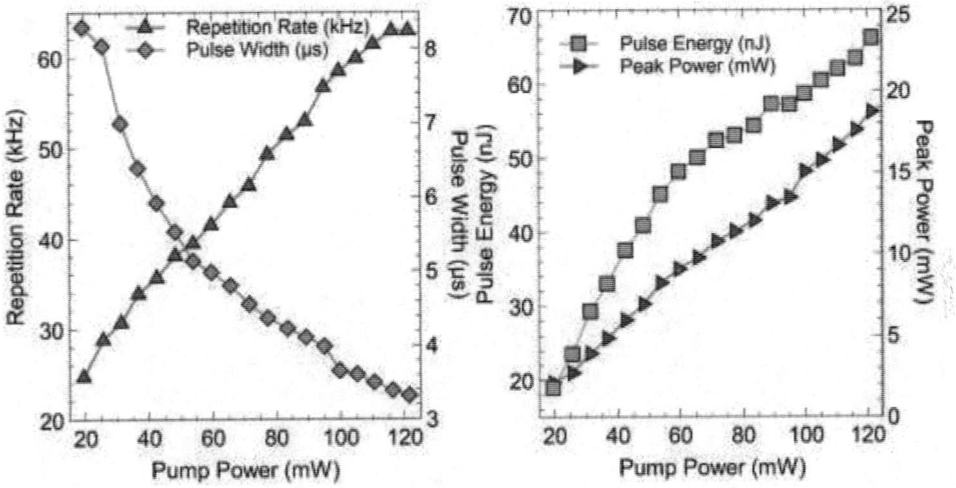

(a) (b)

Figure 3. Relationship between (a) repetition rate and pulse width, and (b) pulse energy and instantaneous peak power at different pump power.

4. Conclusion

In a conclusion, AgNP embedded into the PVA as SA demonstrated the ability to generate Q-switched pulse at an Erbium window. The AgNP was synthesized using the chemical reduction method and then mixed with PVA aqueous solution. The AgNP-PVA was left to dry in the ambient for 72 hours to develop an SA. The proposed experimental works generated pulse energy and peak power of about 66.14 nJ and 18.71 mW. The pump power when increased from 19.4 mW to 121.5 mW, generates a maximum of 62.89 kHz of repetition rate with a shortest pulse width of 3.32 μs.

References

[1] Woodward R I and Kelleher E J 2015 2D Saturable Absorbers for Fibre Lasers *Applied Sciences* **5(4)** 1440-56

[2] Kim K H, Husakou A and Herrmann J 2010 Saturable Absorption in Composites Doped with Metal Nanoparticles *Optics Express* **18(21)** 21918-25

[3] Philip R, Kumar G R, Sandhyarani N and Pradeep T 2000 Picosecond Optical Nonlinearity in Monolayer-protected Gold, Silver, and Gold-Silver Alloy Nanoclusters *Physical Review B.* **62(19)** 13160

[4] Lokman M Q, Shafie S, Shaban S, Ahmad F, Jaafar H, Mohd Rosnan R, Yahaya H and Abdullah S S 2019 Enhancing Photocurrent Performance Based on Photoanode Thickness and Surface Plasmon Resonance using Ag-TiO_2 Nanocomposites in Dye-Sensitized Solar Cells *Materials* **12(13)** 2111

[5] Lokman M Q, Rusdi M F, Rosol A H, Ahmad F, Shafie S, Yahaya H, Rosnan R M, Rahman M A and Harun S W 2021 Synthesis of Silver Nanoparticles Using Chemical Reduction Techniques for Q-switcher at 1.5 μm Region *Optik* **244** 167621

Amplitude modulator of Multi-walled Carbon nanotubes/Chitin in the C-band region

N H Muhamad Apandi[1], H Ahmad[2], M H Ibrahim[3] and F Ahmad[1*]

[1] Malaysia-Japan International Institute of Technology (MJIIT), Universiti Teknologi Malaysia, Jalan Sultan Yahya Petra, 54100 Kuala Lumpur, Malaysia
[2] Photonics Research Centre, University of Malaya, 50603, Kuala Lumpur, Malaysia
[3] School of Electrical Engineering, Universiti Teknologi Malaysia, 81310 Skudai, Johor, Malaysia

[*]Corresponding author's email: fauzan.kl@utm.my

Abstract. In ultrafast all-optical signal processing, the all-optical method is crucial, and all-fiber technique offers a wide range of applications in optical communications. This study investigated the amplitude modulation using multi-walled carbon nanotubes (MWCNTs) embedded into chitin as saturable absorber (SA). The MWCNTs-chitin SA is fabricate using a liquid phase exfoliation method to reduce complexity and produce an excellent material quality. In this paper, an optical amplitude modulator produced a linear region with a regression line of the peak intensity at pump power range from 17.92 mW to 67.92 mW with modulation efficiency of 0.50 dB/mW.

1. Introduction

Optical modulation is a mechanism where data information is load upon the light beam. Modulation is achieved by modifying light beam typically such as direction, amplitude, phase, frequency and polarization state [1]. Due to the recent development of nanotechnology and material science in device fabrication, the modulation speed, modulation range broadening and energy consumption reduction has been improved [2]. 2-dimensional (2D) nanomaterials attract researchers as a promising in acquiring opto-electronic applications since the discovery of graphene in 2004 ago. Some of 2D materials have an excellent optical property like high saturable absorption, quick recovery time, high nonlinearity and broadband absorption thus initiate considerable use in photodetectors, optical modulators and optical transistors [3]. The MWCNTs are essentially rolled graphene, which is composed of carbon atoms that have formed hexagonal networks that eventually form a long cylindrical structure [4]. They possess many advantages in nonlinear optics, high thermal conductivity and high mechanical strength [5-6]. Liu et al. (2013) demonstrated an all-optical modulation by using microfiber graphene at wavelength of 1550 nm with modulation efficiency of 0.002 dB/mW [7]. Ahmad et al. (2019) proposed a heterostructure thin film of molybdenum tungsten disulphide and reduced graphene oxide in polyvinyl alcohol ($MoWS_2$-rGO/PVA) as saturable absorber. The modulation efficiency obtained was 0.30 dB/mW at 1550 nm wavelength [3]. On the other hand, Chong et al. (2016) achieved modulation efficiency of 0.09 nm at wavelength of 1550 nm by integrated optical waveguide with a coating of graphene oxide (GO) [8]. Ahmad et al. (2020) also generated amplitude modulator by employed bismuth telluride (Bi_2Te_3) flakes through mechanical exfoliation. Two values of modulation efficiency acquired were 0.05 dB/mW at pump power range of 0.00 mW to 80.32 mW and 0.81 dB/mW at pump power of 80.32 mW to 98.64 mW, respectively [9].

Through liquid phase exfoliation, SAs from various classes of materials have been fabricated by using two commonly water polymer such as polyvinyl alcohol (PVA) and polyethylene oxide (PEO). In this work, the development of optical modulation utilized MWCNTs embedded in chitin as SA. Chitin is a biopolymer derived from mushrooms [10] was used as an alternative to a man-made polymer. The MWCNTs-chitin were synthesized using liquid exfoliation method due to its simplicity to develop a composite SA. The MWCNTs-chitin SA then integrated into the linear cavity laser in the C-band region.

2. Methodology

The MWCNTs materials were first utilized to fabricate the SA by embedding them in acrylonitrile-butadine-styrene (ABS) 3D printer filament, which can be obtained offline or online; commercially. In this experiment, the diameter and weight of the filament that have been purchased online from 3Dxtech is 1.75 mm and 750 g, respectively. First and foremost, the MWCNTs-ABS printer filament was extruded by utilizing a 3D printer through a nozzle diameter of 0.4 mm at 210 °C. The main aim was to trim the diameter of the filament to 200 μm. Later, the filament was then weighed for at 25 mg and combined with 1 ml of acetone before being sonicated for 5 minutes to dissolve the ABS. Finally, the MWCNTs-acetone suspension is produced. MWCNTs-chitin composite was prepared by adding the MWCNTs-acetone suspension into a chitin with one to one ratio totalling up to 5 ml to produce a film with thickness around 50 μm for a strong and sturdy film that would be easily peeled off later. The MWCNTs-chitin composite was realized by sonification process for more than one hour. The mixture was then placed into a circular shape plastic petri dish with diameter of 32 mm and 10 mm height. It was then was left to dry in ambient temperature for 36 hours to allow ample time for the solution to dry up and thus generating MWCNTs-chitin film SA.

To generate the amplitude modulation, MWCNTs-chitin SA was further cut into small pieces of ~1 $mm^2 \times 1\ mm^2$ and transferred onto a FC/PC fiber end to function as SA. The arrangement of the linear cavity setup used in this experiment was identical to previous work [3] as in Figure 1.

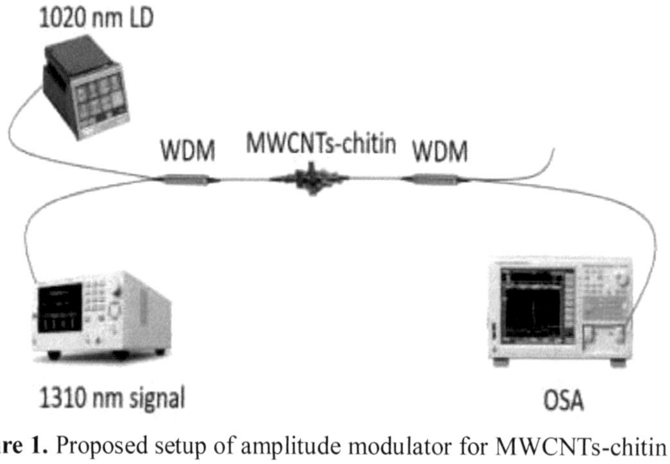

Figure 1. Proposed setup of amplitude modulator for MWCNTs-chitin SA.

3. Results and discussion

Throughout the experiment, the signal light is adjusted at -13.6 dBm or roughly 1.25 mW, resulting in a spectrum peak of -30.12 dBm, as shown in Figure 2. The low signal light guarantees that the 1020 nm pump light induces heat buildup in MWCNTs-chitin SA. The amplitude modulation is evaluated entirely in the spectrum domain to minimize pump light leakage in the WDM, which would impact the modulation results in the time domain if a photodetector and oscilloscope were used. A channel selection or wavelength filter can be added after the second WDM to reduce pumping laser leakage at the output, making the suggested system more feasible in real-world applications. The spectrum is set at 1550 nm at the C-band region. The attenuation trend becomes linear as the pump power is increased further to 17.92 mW, as illustrated in Figure 3 in the linear area between 17.92 mW and 67.92 mW with a modulation efficiency of 0.50 dB/mW.

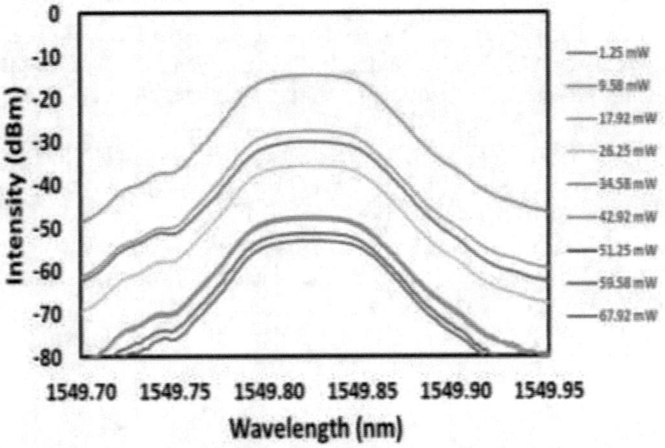

Figure 2. Amplitude modulation of MWCNTs-chitin SA in the optical domain of C-band region.

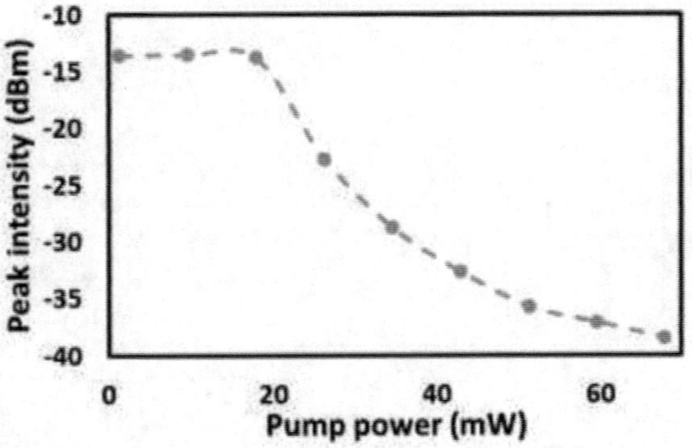

Figure 3. Trend of amplitude modulation of MWCNTs-chitin SA.

4. Conclusion

As a conclusion, we demonstrated MWCNTs-chitin SA has a great potential as an amplitude modulator in the C-band area, with a modulation efficiency of 0.50 dB/mW throughout a pump power range from 17.92 mW to 67.92 mW. Optical modulators play an importance role in numerous fields including optical interconnection, environmental monitoring, biosensing, security applications and medicine. Hence, it is essential to exploit 2D materials for optical modulation to improves the modulation performance such as modulation bandwidth, modulation depth and energy consumption.

References

[1] Dong J, Zhang X. Optical modulators based on 2D materials. In2D Materials for Photonic and Optoelectronic Applications 2020 Jan 1 (pp. 37-77). Woodhead Publishing.

[2] Luo S, Wang Y, Tong X, Wang Z. Graphene-based optical modulators. Nanoscale research letters. 2015 Dec;10(1):1-1.

[3] Ahmad H, Aidit SN, Tiu ZC, Ooi SI. Application of MoWS2-rGO/PVA thin film as all-fiber pulse and amplitude modulators in the O-band region. Optical Fiber Technology. 2019 Mar 1;48:1-6

[4] Zuikafly SNF, Ahmad F, Ibrahim MH, Latif AA, Harun SW. Demonstration of passive saturable absorber by utilizing MWCNT-ABS filament as starting material. InIOP Conference Series: Materials Science and Engineering 2017 Jun 1 (Vol. 210, No. 1, p. 012030). IOP Publishing.

[5] Sobon G. Application of 2D materials to ultrashort laser pulse generation. Two-dimensional Materials-Synthesis, Characterization and Potential Applications. 2016 Aug 31;1.

[6] Han SA, Bhatia R, Kim SW. Synthesis, properties and potential applications of two-dimensional

transition metal dichalcogenides. Nano Convergence. 2015 Dec;2(1):1-4.
[7] Liu ZB, Feng M, Jiang WS, Xin W, Wang P, Sheng QW, Liu YG, Wang DN, Zhou WY, Tian JG. Broadband all-optical modulation using a graphene-covered-microfiber. Laser Physics Letters. 2013 Apr 18;10(6):065901.
[8] Chong WY, Lim WH, Yap YK, Lai CK, De La Rue RM, Ahmad H. Photo-induced reduction of graphene oxide coating on optical waveguide and consequent optical intermodulation. Scientific reports. 2016 Apr 1;6(1):1-8.
[9] Ahmad H, Yusoff N, San Lim H, Mat Jafri MZ, Zulkifli MZ, Tiu ZC. Temporal and amplitude modulation at C-band region using Bi2Te3-based optical modulator. Journal of Modern Optics. 2020 Apr 15;67(7):638-46.
[10] Park BK, Kim MM. Applications of chitin and its derivatives in biological medicine. International journal of molecular sciences. 2010 Dec;11(12):5152-64.

Liquid-Phase Exfoliated Graphene-MoS2 Based Saturable Absorber for Q-switched Erbium Doped Fiber Laser

S. N. M. Halim[1], N. A. H. Jasni[1], M.F.M. Taib[2], W. M. F. W Nawawi[3], F. Ahmad[1]*

[1] Malaysia-Japan International Institute of Technology, Universiti Teknologi Malaysia, Jalan Sultan Yahya Petra, 54100, Kuala Lumpur, Malaysia
[2] Faculty of Applied Sciences, Universiti Teknologi MARA (UiTM), 40450 Shah Alam, Malaysia
[3] Department of Biotechnology Engineering, International Islamic University Malaysia, Kuala Lumpur, Malaysia

*Corresponding author's email: fauzan.kl@utm.my

Abstract. Upon exfoliation from the bulk form, two-dimensional materials have shown ubiquitous properties which are suitable for Q-switched pulsed laser generation. In this research, a successful solution process of graphene-MoS_2 nanocomposite saturable absorber through liquid phase exfoliation has been carried out. The method offers a low-cost route for simple and scalable production while providing a promising material quality with on-demand properties and integration flexibility. Stable Q-switched laser operation was realized with graphene-MoS2 hybrid saturable absorber. The pulse duration was measured to be 6 μs with repetition rate of 63.92 kHz corresponding to a peak power and pulse energy of 5.05 mW and 30.87 nJ, respectively.

1. Introduction

Optical fiber system with broadband and tunable saturable absorber (SA) is important for creating a compact Q-switched fiber laser. The discovery of two-dimensional (2D) graphene offers plenty of opportunities for the next generation of SAs to be used in passive Q-switching [1]. The gapless nature of graphene has led to the wavelength-independent absorption which makes it suitable for broadband spectrum. Inspired by the disclosure of graphene phenomenon, a significant number of other 2D materials such as transition metal dichalcogenides (TMDs), phosphorene, topological insulators, and MXenes have been actively pursued [2].

Not only that, many endeavours have concentrated their works in the avenues of synthesis and viable routes for scalable production and exploration of properties that will enable novel device applications. Liquid-phase exfoliation (LPE) is one of the eminent methods of choice among researchers for 2D materials fabrication. This is due to its simplicity, cost-effectiveness, and high scalability for production of large quantities of mono- and few layer 2D materials in a commercially viable way [3]. Therefore, this research work proposed a simple solution process through liquid-phase exfoliation of graphene-MoS_2 nanocomposite for passive Q-switched pulse laser with chitin as a biocompatible host polymer.

Journal of Physics: Conference Series **2075** (2021) 012006 doi:10.1088/1742-6596/2075/1/012006

2. Methodology

2.1. Graphene-MoS2 SA fabrication
Graphene solution was prepared with conductive graphene-PLA based 3D printer filament as the starting material. The graphene-PLA filament will be extruded through a 3D printer nozzle and weighed to 25 mg before dissolved in 1 ml of THF solution. The graphene-THF suspension is then undergo ultrasonication for 10 minutes. For MoS_2 solution, 5 mg MoS_2 powder and 5 ml THF solution are mixed together until MoS_2 powder dissolved homogeneously in THF solution with stirring time of about 2 hours, at ambient temperature and 1200 rpm. Meanwhile, a host polymer is needed to become a binder in the fabrication of SA film. A recent bio-host polymer namely chitin which is prepared from the oyster mushroom is introduced in this research. The synthesis work for chitin biopolymer can be found in report by Wan Nawawi et al. [4]. Solution-processed of graphene-MoS_2 hybrid is done by mixing the prepared graphene and MoS_2 solution with chitin in one-to-one ratio and placed in ultrasonic bath for sonication process about 60 minutes to ensure a well-homogenized dispersion form with good viscosity matrices. Ample time of sonication process allowed exfoliation of bulk crystals into nanosheets as the van der Waals interaction weaken due to production of microjets and shockwaves from collapsing cavitation bubble. The obtained dispersion mixtures are then poured into 35 mm × 15 mm circular shape petri dish and let dry in ambient temperature for about 36 hours before they ready to be peeled off as a free-standing film SA.

2.2. Experimental setup
The graphene-PLA filament will be extruded through a 3D printer nozzle and weighed to 25 mg before dissolved in 1 ml of THF solution. The fabricated SA was integrated into fiber laser cavity by sandwiching it between two ferrules as shown in Figure 1.

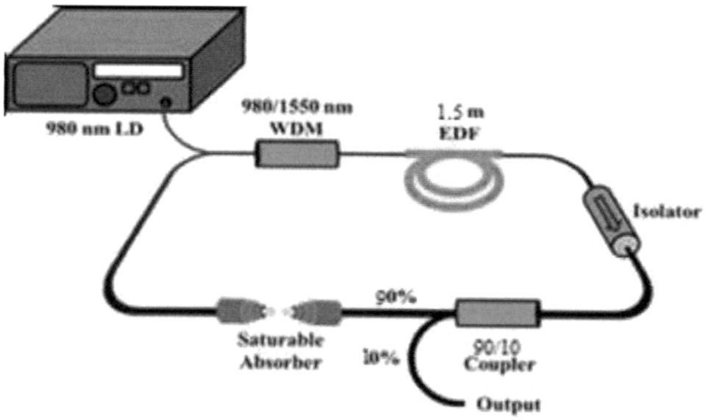

Figure 1. Laser cavity setup of Q-switched EDFL based on graphene-MoS_2 chitin SA.

The output was tapped out by a 90/10 optical coupler, where 90 % of the light is kept oscillating in the cavity, and the remaining 10 % is used for output measurement to simultaneously observe the pulse train, signal to noise ratio and optical spectrum through a 3 dB coupler. The output are obtained through Optical spectrum analyzer (OSA) (Yokogawa AQ6370B), optical power meter (OPM) (Thorlab PM100D), mixed domain oscilloscope (Tektronix MDO3024) which were connected through a 460 kHz bandwidth photodetector (PD) (Thorlab DET01CFC).

3. Results and Discussions

The performance of SAs in generating pulsed laser were graphed for graphene-MoS_2 chitin nanocomposite at one-to-one ratio. Figure 2 shows optical spectrum with Q-switched pulse lasing operated at around 1558 nm at maximum pump power of 104.93 mW when fabricated SA is integrated into the laser cavity.

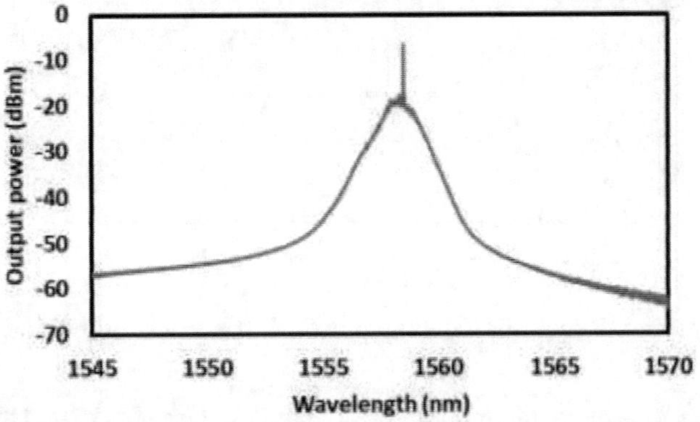

Figure 2. Optical spectrum at maximum pump power of 104.93 mW.

The pulse train repetition rate and the associated pulse width of graphene-MoS_2 is shown in Figure 3 as recorded at maximum input pump power. The graphene-MoS_2-chitin based SA film was able to generate Q-switching pulse pump power of 66.32 mW, tunable up to 104.93 mW. The graph also indicates a highly stable Q-switching operation with little to no amplitude modulation can be obtained from graphene-MoS_2 SA.

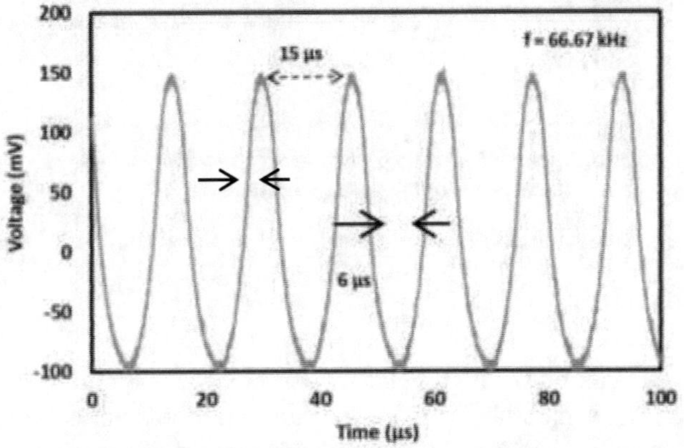

Figure 3. Pulse train repetition rate and pulse width at maximum input pump power.

Journal of Physics: Conference Series **2075** (2021) 012006 doi:10.1088/1742-6596/2075/1/012006

The relationship between repetition rate and pulse width against the input pump power is depicted in Figure 4.

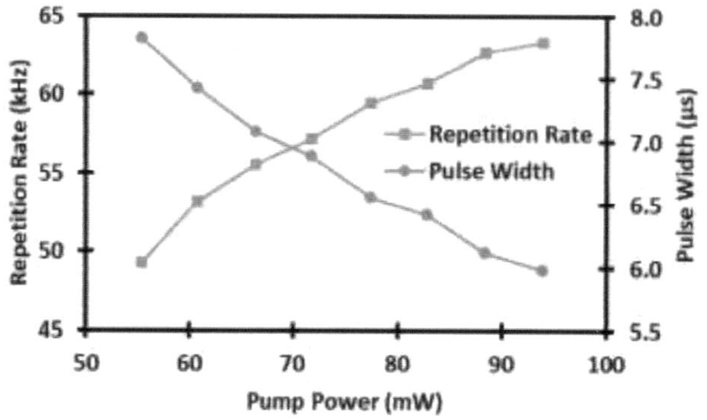

Figure 4. Repetition rate and pulse width against pump power.

As the pump power increased, the repetition rate increased up to 63.29 kHz while the pulse width was reduced to 6 μs. The trend shows a typical Q-switched operation. This resulted in instantaneous peak power and pulse energy of 5.05 mW and 30.87 nJ as shown in Figure 5.

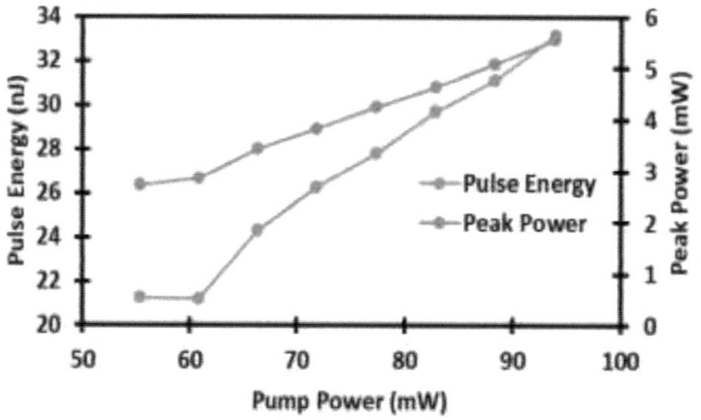

Figure 5. Instantaneous pulse energy and peak power with respect to input pump power.

The pulse energy and peak power is calculated from obtained repetition rates frequency and pulse width. Typically, both of these parameters share the same trend of inclination towards the maximum input pump power. On the other hand, the SA film demonstrated Q-switching pulsed lasing with high stability of over 60 dB. Figure 6 shows the radio frequency spectrum analyser (RFSA) to confirm the stability parameter in pulse laser generation.

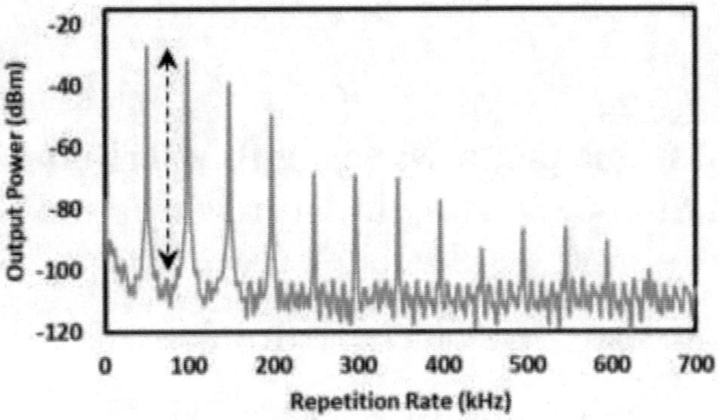

Figure 6. RFSA measurement of the Q-switched EDFL.

4. Conclusion

The SA films of proposed material were successfully fabricated via solution process through liquid phase exfoliation in a desirable condition i.e., scalable production, low-cost consumption, room temperature condition, and minimal apparatus. Apart from that, a passively Q-switched EDFL is successfully demonstrated by using graphene-MoS_2 chitin SA with high repetition rate, low pulse width and good stability and repeatability.

References

[1] Zhang B, Liu J, Wang C, Yang K, Lee C, Zhang H and He J 2020 *Laser & Photonics Reviews* ***14*(2)** 1900240
[2] Ma C, Wang C, Gao B, Adams J, Wu G and Zhang H 2019 *Applied Physics Reviews* ***6*(4)** 041304
[3] Alzakia F I and Tan S C 2021 *Advanced Science* ***8*(11)** 2003864
[4] Fazli Wan Nawawi W M, Lee K Y, Kontturi E, Murphy R J and Bismarck A 2019 *ACS Sustainable Chemistry & Engineering* ***7*(7)** 6492-6496

Generation of stable penta-wavelength ytterbium doped fiber laser by spatial-mode beating in 1-µm waveband

M A Mat Salim[1,2], M A Ismail[3], Saaidal R Azzuhri[4*], M Z A Razak[5], H Bakhtiar[1] and N. Bidin[1]

[1]Laser Center, Ibnu Sina Institute for Scientific and Industrial Research (ISI-SIR), Universiti Teknologi Malaysia 81310 Johor Bahru, Malaysia
[2]T05-Faculty of Science Laboratory Management, Universiti Teknologi Malaysia 81310 Johor Bahru, Malaysia
[3]Research Centre for Photonics and Instrumentation. City University of London, Northampton Square, London EC1V0HB, United Kingdom
[4]Department of Computer Systems and Technology, Faculty of Computer Science and Information Technology, University of Malaya, 50603 Kuala Lumpur. Malaysia
[5]Institute of Microengineering and Nanoelectronics (IMEN), Universiti Kebangsaan Malaysia, 43600 Bangi, Selangor. Malaysia

[*]Corresponding author's email: saaidal@um.edu.my

Abstract. Penta-wavelength ytterbium doped-fiber laser had been successfully generated by usingspatial mode beating method at room temperature. A G652D premium fiber patch cable was inserted inside the polarization controller (PC) and connected to an ytterbium-doped fiber ring cavity. By increasing the pump power, stable multi-wavelength with side-mode suppression ratio (SMSR) from 25 dB to 40 dB were obtained from dual-wavelength and up to penta-wavelengths respectively. The uniform wavelength interval spacing was 0.67 nm whereas peak power fluctuation and wavelength shifts were 0.6 dBm and 0.01 nm, respectively for all sets of multi-wavelength. These simple, easy and reliable system offer a stable multi-wavelength generation in a 1-micrometer waveband.

1. Introduction

Recently, fiber lasers with the multi-wavelength operation have great potential in a multitude of operations for example in the measurement of dense wavelength-division multiplexed (DWDM), fiber optics devices testing, sensor multiplexing, instrumentations, and telecommunications system. Substantial efforts are currently explored to develop stable, dense, and multi-wavelength operations. Multi-wavelength fiber laser (MWFL) based on an erbium-doped fiber (EDF) as an active medium has presented significant advantages and highest potential for further exploration, offering compatibility with existing telecommunication facilities, high power efficiency, and low threshold power compared to other active medium [1]. Despite that, EDF-based MWFL has unstable multi-wavelength operation and mode competition due to cross-saturation gain and strong homogenous broadening at room temperature. Therefore, various research works have been proposed to overcome this situation, i.e. by cooling the EDF in liquid nitrogen temperature [2] however the technique requires costly equipment and complexity. Others proposed polarization hole burning (PHB) effect [3] and non-linear effect such as a combination of stimulated Brillouin scattering with a four-wave mixing [4] and spatial-mode beating [5]. There also those who used fiber bragg grating [6], tapered fiber [7-9], photonics crystal fiber [10-

12], side polished fiber [13] highly nonlinear fiber [14] and arrayed waveguide grating (AWG) [15] to be incorporated into cavity ring for the purpose of generating dual- and MWFL as wavelength selective filter. The simplest and cheapest method to set up a multi-wavelength operation is by using a spatial mode beating in an EDFL [16].

Besides, generation of multi-wavelength laser based on ytterbium doped fiber (YDF) as active medium is much easier due to Yb^{3+} ions have larger inhomogeneity of gain profile and broader spectral range of 970nm – 1.2-micron region. Of these potentials, YDF based MWFL has also less attraction and far less reports as compared to EDF based MWFL. Multi-wavelength is favourable due to its extensive application, for examples optical sensing, wavelength division multiplexed (WDM) communication and field of instrument testing. Therefore, there are opportunities and growing demand for further investigation of the multi-wavelength development in 1μm region. There are few ways can be implemented to induce generation of multi-wavelength in YDFL at room temperature.

In this paper, spatial-mode beating method was applied to achieve multi-wavelength in YDFL. A compact YDF-based MWFL was successfully demonstrated by spatial-mode beating in which 1m of G652D premium fiber path cable was incorporated inside the three-rotatable waveplates of polarization controller (PC) which was then connected into the ring cavity to generate switchable multi-wavelength fiber laser in the region of 1 μm waveband. The uniform interval wavelength spacing is 0.67 nm from dual-wavelength up to penta-wavelength with wavelength shift and peak power fluctuation were 0.01 nm and 0.6 dBm, respectively.

2. Methodology

The experimental setup for penta-wavelength fiber laser is depicted in Figure 1, which comprise of 70cm length of Ytterbium-doped fiber (YDF) as the active medium and a G652D premium fiber patch cable with FC/PC terminated fiber as a wavelength-selective filter in a ring cavity fiber laser. The YDF has peak absorption of 1300 dBm at 977 nm. It was pumped by a 974nm laser diode which was multiplexed using a 980/1060 nm wavelength division multiplexer (WDM). A polarization insensitive isolator was connected to ensure the propagation of laser in a unidirectional manner and for this work, clockwise direction. A polarization controller (PC) with three-rotatable waveplates was used in the laser cavity to control and optimized the birefringence. The other end of the PC was attached to a 90:10 optical coupler (OC) where 10% of the laser output was extracted and routed to an optical spectrum analyzer (OSA) with 0.02 nm spectral resolution.

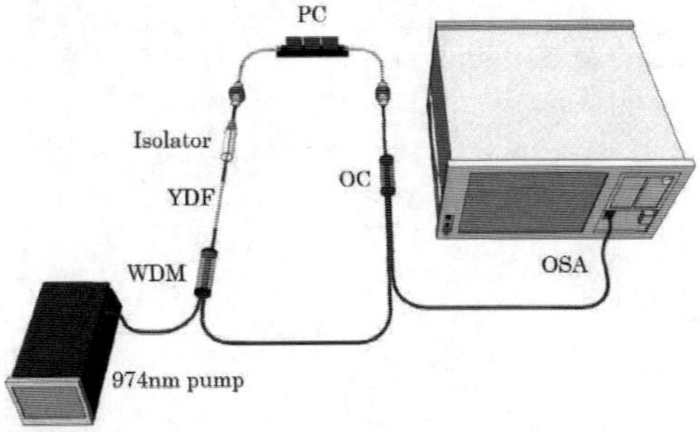

Figure 1. Experimental setup of penta-wavelength fiber laser generation in 1μm waveband

A 1-meter single mode fiber G652D premium fiber patch cable was inserted into the polarization controller (PC) and connected to an ytterbium-doped fiber ring cavity by using FC fiber optic adapter which connect two optical cables together. The PC has three-rotatable waveplates which allow us to insert a single mode fiber FC/PC G652D premium fiber patch cable by wrapping the fiber around two or three spools to create independence wave plates that will alter the polarization of transmitted light inside the single mode fiber. The single mode G652D fiber whose wavelength range from 1260 nm to

1625 nm, is used as 'multimode' fiber for fiber laser. The coupled fiber between single mode-multimode-single mode, whose excited LP_{01} and LP_{11} is used as spatial-mode beating filter to generate multi-wavelength fiber laser. In this way, at first connection between single mode-multimode, diffracted the light into a few waveguide modes especially LP_{01} and LP_{11}. Generally, both LP_{01} and LP_{11} propagate inside the multimode section with differential constant;

$$\Delta\beta = \beta(LP_{01}) - \beta(LP_{11}) \quad (1)$$

with beat length given by;

$$L_B = 2\pi/\Delta\beta. \quad (2)$$

The second connection between multimode-single mode, the diffraction of the light combined back into the single mode waveguide. If the length of multimode fiber is larger than spatial mode beating length, then the comb spectrum will generate and lasing wavelength arises at the transmission peak. The wavelength spacing of the lasing wavelengths depends on the propagation constants different and multimode fiber length.

3. Results and Discussion

The lasing threshold for Ytterbium-doped fiber laser (YDFL) started at 48.5 mW with wavelength and peak power of 1067.50nm and -33.9 dBm, respectively. The single lasing spectrum remains constant, however its peak power increased to -20.5 dBm as the pump power was further increased up to 74.8 mW to achieve stable operation in ring cavity configuration.

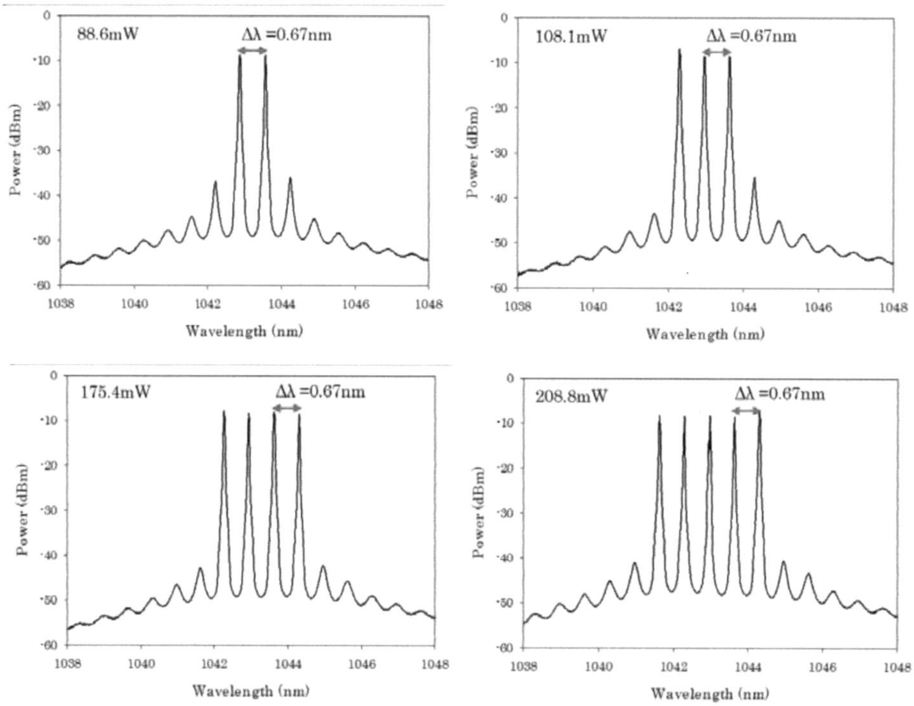

Figure 2. Generation of dual-wavelength, triple-wavelength, quad-wavelength and penta-wavelength at 88.6mW, 108.1mW, 175.4mW and 208.8mW respectively.

In this condition, PC with three-rotatable waveplates was in fixed position. As pump power was increased to 88.6mW, dual-lasing wavelength obtained at 1042.89 nm and 1043.56 nm with peak power of -7.81 dBm and -7.93 dBm respectively. Then, by increasing pump power to 108.1mW triple-lasing wavelengths obtained at 1042.27 nm, 1042.94 nm and 1043.61 nm with peak power of -7.00 dBm, -6.99 dBm and -7.67 dBm respectively. Both dual and triple lasing wavelengths have a uniform wavelength spacing of 0.67nm whereas side-mode suppression ration (SMSR) was about 25 dB and 35 dB respectively. When pump power was further increased to 175.4 mW, quad-lasing wavelength obtained at 1042.27 nm, 1042.94 nm, 1043.61 nm and 1044.28 nm with with peak power of -8.01 dBm, -8.52 dBm, -7.99 dBm and -8.53 dBm respectively. Its wavelength spacing remains uniform with 0.67 nm and SMSR was about 35 dB. As the pump power was increased at the highest level, penta-lasing wavelength obtained at 1041.60 nm, 1042.27 nm, 1042.94 nm, 1043.61 nm and 1044.28 nm with peak power of -8.34 dBm, -8.56 dBm, -8.82 dBm and -7.40 dBm respectively. The generated penta-lasing wavelength were located in the range from 1041.60 nm to 1044.28 nm with a uniform wavelength spacing of 0.67 nm and SMSR of 40 dB.

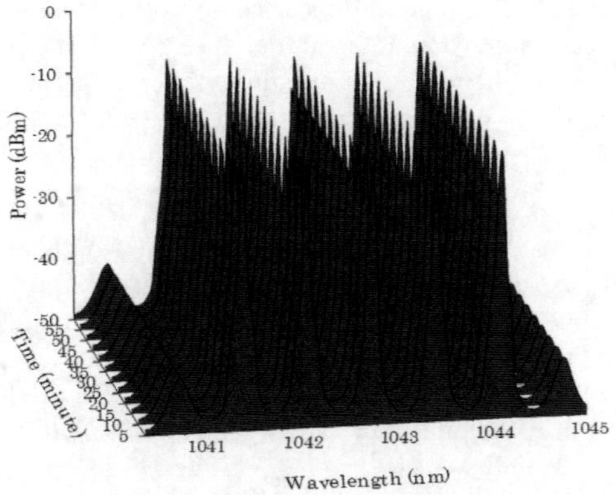

Figure 3. Stability of penta-wavelength YDFL with a uniform wavelength spacing.

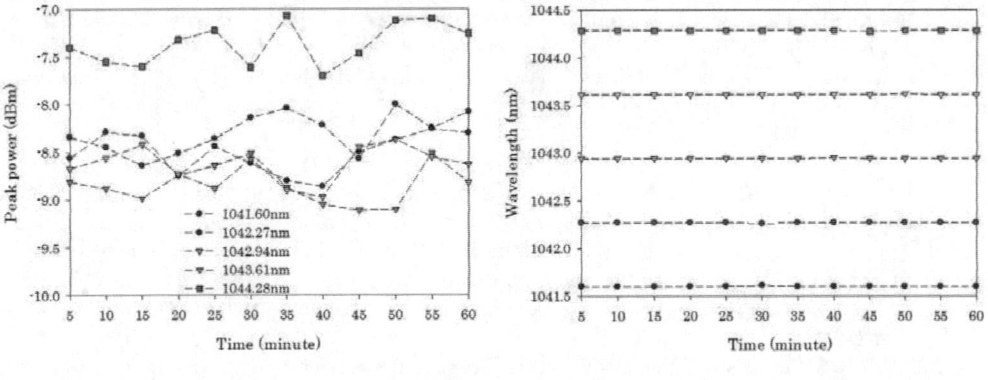

(a) (b)

Figure 4. (a) Power fluctuation and (b) wavelength shifting of penta-wavelength YDFL

The penta-wavelength fiber laser has a uniform wavelength spacing and stable peak power fluctuation for an hour of operation at room temperature as depicted in Figure 3. The peak power of lasing wavelength was observed very stable over time as shown in Figure 4(a) with minimal fluctuation of 0.6 dB. Figure 4(b) show wavelength shifting of 0.01 nm for generated penta-wavelength fiber laser. In addition, instead of increasing pump power, by adjusting the PC will also generate number of lasing wavelengths. Furthermore, its lead to relatively uniform peak power fluctuation, wavelength interval spacing and stability of penta-wavelength fiber laser over time. The multimode PC allow polarization adjustment of spatial mode in the ring cavity and permit energy interchange between lasing modes thus distributing evenly over lasing wavelengths [5].

4. Conclusion

A stable and compact Ytterbium-doped fiber based multi-wavelength fiber lasers are successfully demonstrated by inserting a G652D premium fiber patch cable inside the PC and then connected into the YDFL cavity rings configuration for 1-micron spectral region. The proposed multi-wavelength with a G652D fiber patch cable acts as a multimode fiber is capable to generate stable dual, triple, quad, and penta-wavelength with wavelength shift and minimal peak power fluctuation of 0.01 nm and ~0.6 dB, respectively over an hour operation at room temperature. The SMSR and interval wavelength spacing are 40 dB and 0.67 nm respectively. In addition, this configuration is simple and easy to integrate into a fiber laser system, compact, and also offers a promising alternative for MWFL.

5. References

[1] K. K. Qureshi, H. Y. Tam, W. Chung, and P. K. A. Wai, "Multiwavelength laser source using linear optical amplifier," *Ieee Photonic Tech L,* vol. 17, no. 8, pp. 1611-1613, 2005.

[2] S. Yamashita and K. Hotate, "Multiwavelength erbium-doped fibre laser using intracavity etalon and cooled by liquid nitrogen," *Electronics Letters,* vol. 32, no. 14, pp. 1298-1299, 1996.

[3] S. Pan, & J. A Yao, "Wavelength-switchable single-longitudinal-mode dual-wavelength erbium-doped fiber laser for switchable microwave generation". *Optics express*, *17*(7), 5414-5419 2009.

[4] Y. Yuan, M. Yi, and Y. Yao, "Multi-band multiwavelength fiber laser based on the combined effects of four wave mixing and self-seeding stimulated Brillouin scattering," *Optics Communications,* vol. 395, pp. 88-93, 2017.

[5] Z. Li, C. Lou, and Y. Gao, "A multiwavelength erbium-doped fiber laser realized by controlling the polarization state," *International journal of infrared and millimeter waves,* vol. 20, no. 8, pp. 1487-1491, 1999.

[6] L. Zhang *et al.*, "Six-wavelength-switchable narrow-linewidth thulium-doped fiber laser with polarization-maintaining sampled fiber Bragg grating," *Optics & Laser Technology,* vol. 136, p. 106788, 2021.

[7] H. Ahmad, M.A.M. Salim, S. Azzuhri, M. Zulkifli, and S. Harun, "Dual wavelength single longitudinal mode Ytterbium-doped fiber laser using a dual-tapered Mach-Zehnder interferometer," *Journal of the European Optical Society-Rapid publications,* vol. 10, 2015.

[8] H. Ahmad, M. A. M. Salim, S. R. Azzuhri, and S. W. Harun, "Highly stable and tunable narrow-spacing dual-wavelength ytterbium-doped fiber using a microfiber Mach-Zehnder interferometer," (in English), *Optical Engineering,* vol. 55, no. 2, p. 026114, Feb 2016, doi: Artn 026114

[9] M. A. M. Salim *et al.*, "Switchable erbium-doped fiber laser (EDFL) using non-adiabatic microfiber," *Optical Fiber Technology,* vol. 52, p. 101967, 2019.

[10] H. Ahmad, M.A.M. Salim, S. R. Azzuhri, M. Soltanian, S. Harun, and M. Yasin, "Generation of an ultra-stable dual-wavelength ytterbium-doped fiber laser using a photonic crystal fiber," *Laser Physics,* vol. 26, no. 2, p. 025101, 2015.

[11] M.A.M. Salim, R. Shaharuddin, M. Ismail, S. Harun, H. Ahmad, and S. R. Azzuhri, "Bi2Te3 based passively Q-switched at 1042.76 and 1047 nm wavelength," *Laser Physics,* vol. 27, no. 12, p. 125102, 2017

12] H. Ahmad, M.A.M. Salim, S. R. Azzuhri, M. Soltanian, S. Harun, and M. Yasin, "Generation of an ultra-stable dual-wavelength ytterbium-doped fiber laser using a photonic crystal fiber," *Laser Physics,* vol. 26, no. 2, p. 025101, 2015.

[13] M. Rahman, F. Rashid, H. Ahmad, S. R. Azzuhri, M. Razak, and M. Faruki, "Dual-wavelength ytterbium-doped fiber laser using microfiber and D-shaped polished fiber," *Optik,* vol. 130, pp. 1421-1425, 2017

[14] M. A. M. Salim, S. R. Azzuhri, M. A. Khudus, M. Razak, N. Nasir, and I. Amiri, "Generation of dual-wavelength ytterbium-doped fibre laser using a highly nonlinear fibre," *Laser Physics,* vol. 28, no. 11, p. 115107, 2018.

[15] M. A. M. Salim, H. Ahmad, S. R. Azzuhri, M. Z. A Razak, & S. W. Harun, "Tunable wavelength generation in the 1 μm region incorporating a 16-channel arrayed waveguide grating (AWG) ". *Laser Physics*, *27*(12), 125101, (2017)

[16] R. Chi, K. Lu, and S. Chen, "Multi-wavelength Yb-doped fiber-ring laser," *Microwave and Optical Technology Letters,* vol. 36, no. 3, pp. 170-172, 2003.

Acknowledgement

The authors would like to express their gratitude to the Ministry of Higher Education Malaysia (MOHE); University of Malaya, Universiti Kebangsaan Malaysia, and Universiti Teknologi Malaysia for the financial support through grant number IIRG028C-2019, FRGS/1/2020/STG07/UKM/02/3, and R.J130000.7809.4L671 respectively.

Generation of stable mode-locked fiber laser based MWCNTs in 1.5-µm waveband

M A Mat Salim[1,2], M A Ismail[3], M Z A Razak[4*], Saaidal R. Azzuhri[5], H Bakhtiar[1] and N Bidin[1]

[1]Laser Center, Ibnu Sina Institute for Scientific and Industrial Research (ISI-SIR), Universiti Teknologi Malaysia 81310 Johor Bahru, Malaysia
[2]T05-Faculty of Science Laboratory Management Centre, Universiti Teknologi Malaysia 81310 Johor Bahru, Malaysia
[3]Research Centre for Photonics and Instrumentation. City University of London, Northampton Square, London EC1V0HB, United Kingdom
[4]Institute of Microengineering and Nanoelectronics (IMEN), Universiti Kebangsaan Malaysia, 43600 Bangi, Selangor. Malaysia
[5]Department of Computer Systems and Technology, Faculty of Computer Science and Information Technology, University of Malaya, 50603 Kuala Lumpur. Malaysia

[*]Corresponding author's email: zul.hakimi@ukm.edu.my

Abstract. Pulses fiber laser had been successfully generated by using multi-wall carbon nanotubes thin film saturable absorber at room temperature. The saturable absorber is incorporated into a ring laser cavity. A stable Kelly sidebands mode-locked pulse spectrum with 1561.3 nm wavelength at the pump power of 86.8 mW. The repetition rate and pulse width of 12.3 MHz and 0.51 picosecond, respectively. A stable operation is observed for an hour at room temperature. These simple and reliable system features offer interesting research study especially in mode-locked pulse generation at 1.5 µm waveband.

1. Introduction

A few years ago, mode-locked fiber lasers (MLFL) had great potential in a multitude of applications such as in material micromachining, terahertz spectroscopy, ophthalmology, space ranging, medical imaging, and optical communication. Substantial efforts have been explored to develop a stable, cost-effective design, high-quality beam, compact and alignment-free laser. MLFL based on erbium-doped fiber (EDF) as an active medium has presented significant advantages and the highest potential for further exploration, especially passive techniques by incorporating a saturable absorber (SA). A passive technique loss modulation is much shorter compared to an active technique. Due to this advantage, many studies have been conducted on EDF-based MLFL by utilizing carbon-based material as SA, for example, carbon nanotubes [1] and graphene [2]. Carbon nanotubes (CNTs) have been chosen as a suitable saturable absorber due to their benefits such as saturation with high-intensity light, very fast recovery time (1-picosecond), efficient light absorber and emitter, broadband operating bandwidth, moderate damage threshold, and low-cost fabrication process.

CNTs typically have two types of carbon wall structure which are single-wall carbon nanotubes (SWCNTs) and multi-wall carbon nanotube (MWCNTs). MWCNTs [3] also have capabilities as higher thermal properties and ease to fabricate, thus attracted great interest and have been focused on by researchers. MWCNTs form a special class of nanotubes due to their morphology and characteristics are similar to SWCNTs however with more resistance to the chemical. The advantageous of MWCNTs over SWCNTs are their mechanical strength which makes it a strong and sturdy material in term of tensile strength and elastic modulus. In addition,

MWCNTs also have the capability to absorb more photons energy per nanotubes [4]. These properties are due to MWCNTs consists of multiple rolled layers of graphene. In the case of MWCNTs, the only outer layer is modified while protecting the inner layer from oxidation or damage, thus having a higher damage threshold compared to SWCNTs. Therefore, MWCNTs have been proposed to be utilized in cavity rings in order to achieve pulse fiber laser.

Recently, the emergence of other 2D materials have been explored and have successfully generated pulse fiber laser, for example, Bismuth selenide [5, 6], Bismuth telluride [7, 8], Molybdenum disulfide [9], Molybdenum diselenide [10], black phosphorus [11, 12], titanium dioxide [13] and zinc oxide [14]. All these materials have remarkable properties, which makes them a suitable material as a SA. Although with the dumping of new materials, MWCNTs still can be idealized as a suitable saturable absorber and still relevant to generate mode-locked and Q-switched pulse.

In this paper, a compact EDF-based MLFL is successfully achieved by using MWCNT as a SA. The MWCNTs SA is used into the ring cavity to achieve a stable MLFL in the region of a 1.5 μm waveband. The broad bandwidth of the laser spectrum is centered at 1561.3 nm with short pulse duration and repetition rate of 0.51 picosecond and 12.3 MHz, respectively.

2. Methodology

The experimental setup for multi-wall carbon nanotubes based MLFL is depicted in Figure 1, which comprise 60cm of EDF as the gain medium, a single-mode fiber (SMF) with a length of 500 m, and a multi-wall carbon nanotube saturable absorber sandwiched between two fiber ferrules in a ring cavity fiber laser. The EDF is pumped by a 974nm laser diode which had to be attached to a 980/1550 nm wavelength division multiplexer (WDM). An insensitive polarization isolator is connected to ensure the propagation of the laser in unidirectional, and for this work, in a clockwise direction. A polarization controller (PC) is used in the cavity to optimize the birefringence. The other end of the PC is linked to a MWCNT/PVA-based SA and then connected to a 95:5 optical coupler (OC1), where 5% of OC1 port is extracted to another optical coupler (OC2) and then connected to measurement equipment. An oscilloscope, a radio frequency spectrum analyzer (RFSA), an optical spectrum analyzer (OSA) with 0.02 nm spectral resolution and an autocorrelator are used for characterizing the generated pulse fiber laser from the cavity. The 95% of the OC1 port is attached to a SMF and then back to another end of WDM in order to complete the ring cavity configuration.

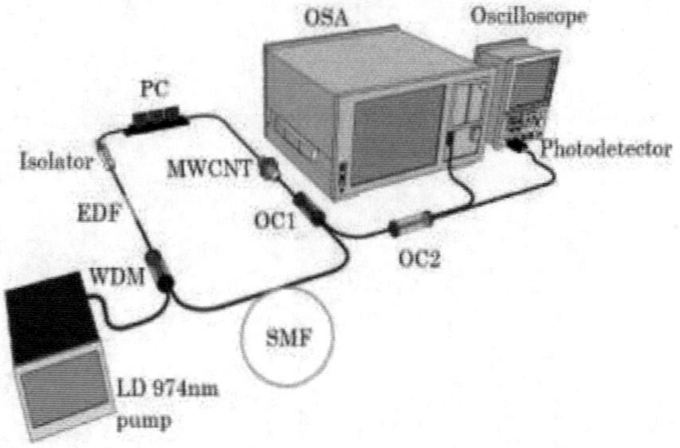

Figure 1. The proposed MWCNTs based mode-locked fiber laser generation in 1.5 μm waveband

The SA is formed by MWCNTs polymer composite thin film with polyvinyl alcohol (PVA). The fabrication process is reported in [15]. The SA has a length of 1–3μm, thickness of about 50μm and diameter range from 10 μm to 20 μm.

3. Results and Discussion

The lasing threshold for EDFL is at 48.5 mW with central wavelength and peak power of 1560.50nm and -33.9 dBm, respectively. The lasing wavelength remains stable by increasing the pump power up to 74.8mW. As the pump power is increased to 86.8mW, the soliton mode-locking has self-started to generate in the ring cavity. The broad mode-locked spectrum is centered at 1561.3 nm with a power level of -32.74 dBm. The optical spectrum is depicted in Figure 2 with the Kelly sidebands spectrum indicated the total cavity dispersion of the pulse in an anomalous dispersion with a total cavity length of approximately 504 m. The broadening of the MLFL sprectum is because of the dispersion of 500 m SMF. The oscillation pulse train of MLFL is shown in Figure 3. Mode-locked pulse with 80.90 ns time interval between pulses can be translated to 12.3 MHz repetition rate, which corresponding to 504 m cavity length. As we decreased or increased the pump power, nothing changes in the repetition rate of the mode-locked pulse. This behaviour proved that the generated pulse train is corresponding to the mode-locking process and is not Q-switched. This is due to the repetition rate of Q-switched changes with the changes of pump power. The generated mode-locked pulse is also measured in repetition rate by radio frequency spectrum analyzer (RFSA) and pulse width by autocorrelator.

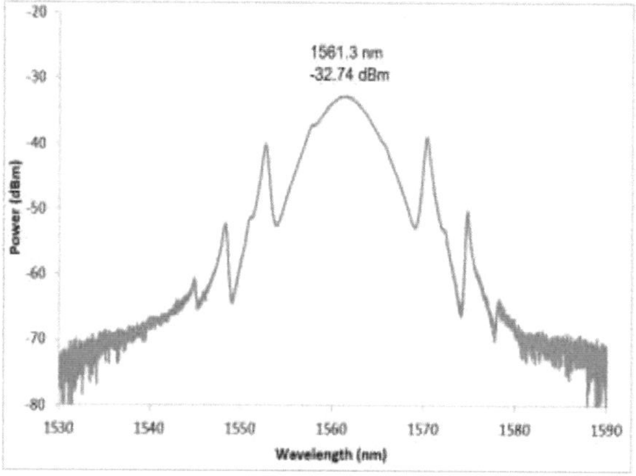

Figure 2. The generation of soliton mode-locking with anomalous dispersion at 1561.3 nm wavelength

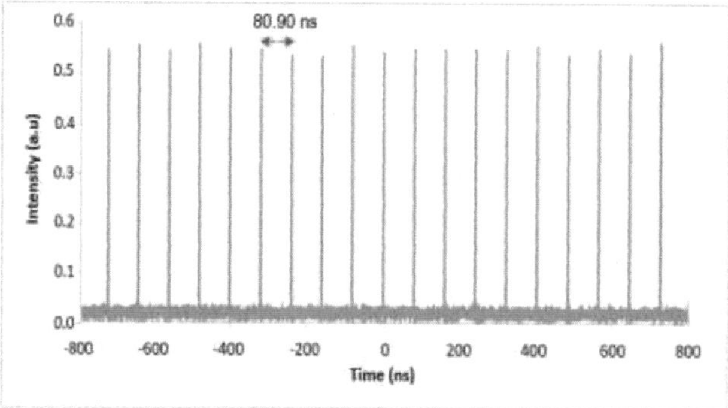

Figure 3. Oscillation of MLFL at 12.3 MHz

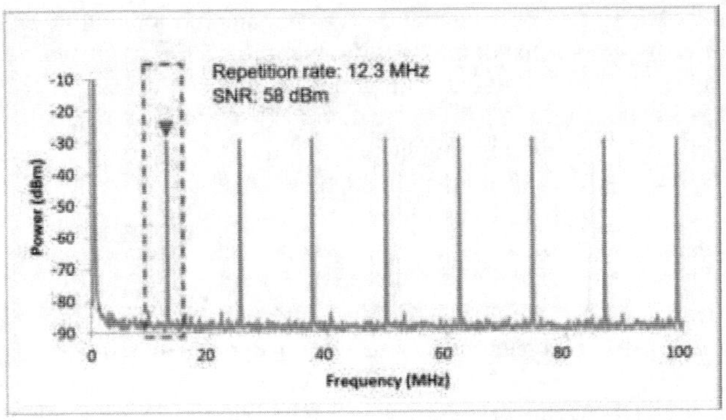

Figure 4. Frequency domain of MLFL at 12.3 MHz

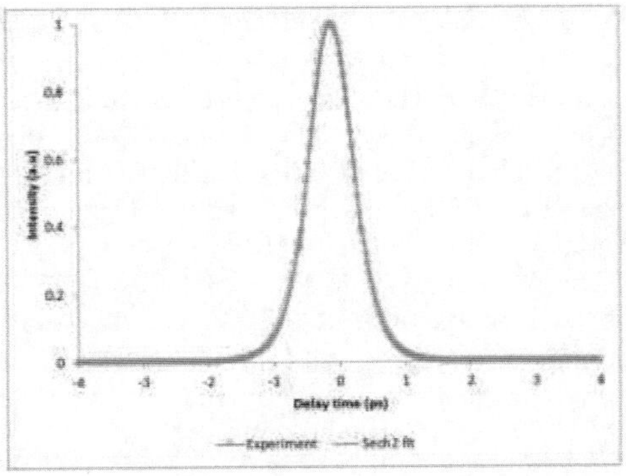

Figure 5. An ultrashort pulse width of MLFL at 0.51 ps

Figure 4 shows the result of the radio frequency spectrum for MWCNTs based mode-locked EDFL with a fundamental harmonic frequency of 12.3 MHz. The signal-to-noise ratio (SNR) is about 58 dB which show the stable operation of the MLFL has been generated in the ring cavity. In addition, no spectral modulation observed in RFSA thus also proved the stability of generated mode-locked pulse. In fact, the MLFL operating signal can last for several days without interference or losing the mode-locking pulse. The stability is also attributed to its remarkable thermal conductivity, where our MWCNTs based SA have capabilities as higher damage threshold. The pulse width of MLFL is about 0.51 picosecond measured by using an autocorrelator. This indicates that the generated pulse occurs at a very fast rate and, to be exact, its ultrashort pulse.

4. Conclusion

A stable and compact mode-locked EDFL is successfully achieved by sandwiching MWCNTs SA between the fiber ferrules in ring cavity configuration for 1.5μm spectral regions. The proposed mode-locked with a MWCNTs acts as a SA is capable of generating a stable mode-locked pulse with a broadband spectrum centered at 1561.3 nm and is observed in a stable condition for over an hour at room temperature. The pulse train repetition rate and pulse widths were 12.3 MHz and 0.51 picoseconds, respectively. In addition, this configuration is relatively simple, and compact demonstrate a promising passive technique to generate mode-locked pulse fiber laser.

5. References

[1] M. Salim, H. Ahmad, S. Harun, N. Bidin, and G. Krishnan, "Generation of an ultrafast femtosecond soliton fiber laser by carbon nanotube as saturable absorber," in *Journal of Physics: Conference Series*, 2018, vol. 1027, no. 1: IOP Publishing, p. 012011.

[2] M. A. M. Salim, M. A. Ismail, M. Z. A. Razak, and S. R. Azzuhri, "Generation of Ultrafast Erbium-Doped Fiber Laser (EDFL) utilizing Graphene Thin Film," in *Journal of Physics: Conference Series*, 2020, vol. 1484, no. 1: IOP Publishing, p. 012026.

[3] K.-N. Cheng, Y.-H. Lin, and G.-R. Lin, "Single-and double-walled carbon nanotube-based saturable absorbers for passive mode-locking of an erbium-doped fiber laser," *Laser Physics,* vol. 23, no. 4, p. 045105, 2013.

[4] Z. C. Tiu, F. Ahmad, S. J. Tan, A. Zarei, H. Ahmad, and S. W. Harun, "Multi-wavelength Q-switched Erbium-doped fiber laser with photonic crystal fiber and multi-walled carbon nanotubes," *Journal of Modern Optics,* vol. 61, no. 14, pp. 1133-1139, 2014.

[5] H. Ahmad, M. Salim, S. R. Azzuhri, M. Soltanian, and S. Harun, "A passively Q-switched ytterbium-doped fiber laser based on a few-layer Bi2Se3 saturable absorber," *Laser Physics,* vol. 25, no. 6, p. 065102, 2015.

[6] H. Ahmad, M. Salim, M. Soltanian, S. R. Azzuhri, and S. Harun, "Passively dual-wavelength Q-switched ytterbium doped fiber laser using Selenium Bismuth as saturable absorber," *Journal of Modern Optics,* vol. 62, no. 19, pp. 1550-1554, 2015.

[7] M. Salim, R. Shaharuddin, M. Ismail, S. Harun, H. Ahmad, and S. R. Azzuhri, "Bi2Te3 based passively Q-switched at 1042.76 and 1047 nm wavelength," *Laser Physics,* vol. 27, no. 12, p. 125102, 2017.

[8] M. A. M. Salim, M. Z. Ab Razak, S. R. Azzuhri, M. A. Ismail, F. Ahmad, and S. W. Harun, "Generation of Microsecond Ytterbium-Doped Fibre Laser Pulses using Bismuth Telluride Thin Film as Saturable Absorber," *Sains Malaysiana,* vol. 48, no. 6, pp. 1289-1294, 2019.

[9] H. Ahmad, Z. C. Tiu, A. Zarei, M. Suthaskumar, M. A. M. Salim, and S. W. Harun, "Domain-wall dark pulse generation in fiber laser incorporating MoS2," *Applied Physics B,* vol. 122, no. 4, pp. 1-5, 2016.

[10] M. A. M. Salim, S. R. Azzuhri, M. A. Ismail, and M. Z. A. Razak, "Few Layer Molybdenum Selenide Saturable Absorber using Optical Deposition Technique for Q-switched Ytterbium Pulses Laser Generation," in *Journal of Physics: Conference Series*, 2020, vol. 1484, no. 1: IOP Publishing, p. 012025.

[11] H. Ahmad, M. Salim, K. Thambiratnam, S. Norizan, and S. Harun, "A black phosphorus-based tunable Q-switched ytterbium fiber laser," *Laser Physics Letters,* vol. 13, no. 9, p. 095103, 2016.

[12] F. Rashid *et al.*, "Using a black phosphorus saturable absorber to generate dual wavelengths in a Q-switched ytterbium-doped fiber laser," *Laser Physics Letters,* vol. 13, no. 8, p. 085102, 2016.

[13] H. Ahmad and M. Salim, "TiO 2-Based Q-Switched Ytterbium-Doped Fiber Laser," *IEEE Journal of Quantum Electronics,* vol. 53, no. 5, pp. 1-6, 2017.

[14] H. Ahmad, M. Salim, M. F. Ismail, and S. W. Harun, "Q-switched ytterbium-doped fiber laser with zinc oxide based saturable absorber," *Laser Physics,* vol. 26, no. 11, p. 115107, 2016.

[15] H. Ahmad, M. Ismail, S. Hassan, F. Ahmad, M. Zulkifli, and S. Harun, "Multiwall carbon nanotube polyvinyl alcohol-based saturable absorber in passively Q-switched fiber laser," *Applied optics,* vol. 53, no. 30, pp. 7025-7029, 2014.

Acknowledgment

The authors are grateful to the Ministry of Higher Education Malaysia (MOHE), Universiti Kebangsaan Malaysia, University of Malaya, and Universiti Teknologi Malaysia for the financial support through grant number FRGS/1/2020/STG07/UKM/02/3, IIRG028C-2019 and R.J130000.7809.4L671 respectively.

Overview of surface plasmon resonance optical sensors for Covid-19 (SARS-CoV-2) detection

H A Zain[1], **M Batumalay**[2*] **and S W Harun**[1]

[1] Department of Electrical Engineering, Faculty of Engineering, University of Malaya, 50603 Kuala Lumpur, Malaysia
[2] Faculty of IT, INTI International University, 71800 Nilai, Negeri Sembilan, Malaysia.

*Corresponding author's email: malathy.batumalay@newinti.edu.my

Abstract. The SARS-CoV-2 i.e., the novel severe acute respiratory syndrome corona virus; has caused massive loss of life. Mitigating this pandemic requires rapid inexpensive technologies for testing COVID-19. Optical sensors can be used to detect the Covid-19 virus by the surface Plasmon resonance phenomenon. Surface plasmon resonance sensors have good sensitivity, response times, fine resolution, and limits of detection. This paper, provides a brief overview on the COVID-19 effects, currently used testing technology, and potential of surface plasmon resonance optical sensors use for detecting this virus.

1. Introduction

The corona virus disease (Covid-19) caused a massive global pandemic and have caused a grave loss of lives. COVID-19 has various symptoms making conclusive diagnosis challenging. Covid-19 symptoms are high fever, dry coughing, sore throat, temporary loss of taste and smell, diarrhea, strong headaches, and sudden fatigue, among others[1]. So, conclusive testing methods are needed to stop this pandemic.

The long waiting lines for Covid-19 testing were caused by the lack of equipment for testing technology COVID-19 are a big concern. The current technologies for Covid-19 testing can be described as slow and expensive. For instance, single photon emission computed tomography (SPECT), computed tomography (CTes), positron emission tomography (PET), enzyme-linked immunosorbent assay (ELISA) can be expensive and laborious procedures and they require well-trained staff to conduct them. The more common testing technology reverse-transcription polymerase chain reaction (RT-PCR) is not cost-effective and has long response time. Thus, RT-PCR testing clinics have been overwhelmed with potential patients. Also, RT-PCR testing requires trained professionals, in some cases has been shown to give false-negatives and can be expensive for the public. [2-6].

Optical sensors have been widely used in many applications due to their higher sensitivity, good selectivity, relatively short response times, and good resolution. [7]. Optical surface plasmon resonance sensors using Kretschmann setup coated with graphene and thiol DNA had a sensitivity of 130 degree/RIU in simulations[8]. Another chip optical sensor with spike protein specific Plasmonic chip had a limit of detection of up to 370 vp/mL [9].Optical sensors can provide results at a much faster times than current technologies with high sensitivity and good limits of detections.

2. Methodology

SPR Kretschmann setup is a surface plasmon sensor structure that has a prism with a very thin metal layer coating. Gold is commonly selected as an SPR coating for good sensitivity and durability. P-polarized (TM) light from a He-Ne laser hits the prism. Then, the light is received by a photo detector. The evanescent field of the light at the prism boundary hits the thin metal layer. when resonance conditions are fulfilled, i.e., P-polarized light, a prism with high refractive index, and a thin uniform metal layer coated on one face of the prism.

Resonance occurs at the surface plasmon resonance angle and the resulting resonance dip can be seen in the output [8, 10]. Surface plasmon resonance sensors need a layer of immobilized biomolecules (such as antibodies) immobilized on the thin metal layer of the prism outer surface. This immobilized bio-molecules layer can bind the pathogen (the virus) to the sensors' surface [11]. The resulting accumulation of the bound pathogen (virus) cause a change (an increase) in the refractive index of the sensor's surroundings. Thus, shifting the sensor resonance angle. [12-14]. By recording the shifts of the resonance angle, the pathogen can be detected.

The SPR sensor setup consists of three main components: the laser light source, the prism, and the light detector. The He-Ne laser can be used as light source [15]. The light detector receives the light from the prism. The experiment setup is presented in Figure 1.

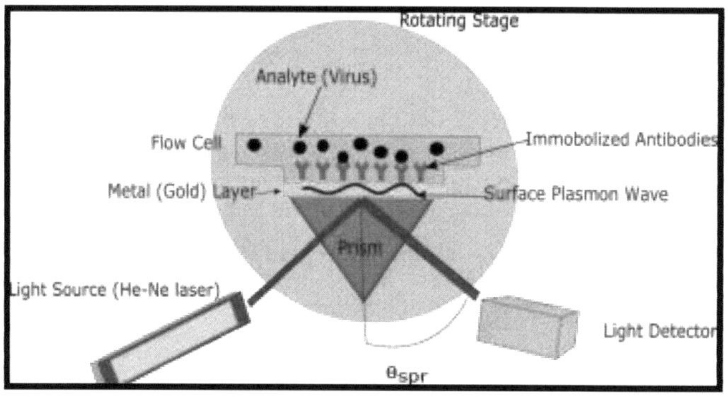

Figure 1. The experiments setup for the pathogen.

3. Results and discussion

Sensitivity, response time, figure of merit, and limit of detection are performance parameters showing the feasibility of using a biosensor. The SPR sensors' sensitivity is the amount of the shift in the resonance angle divided by the total changes of surroundings' refractive index. The binding of the pathogen to the layer of the immobilized ligands on the thin metal coating of the prism surface cause a change of the refractive index.

The centre resonance angle of SPR is the angle of the minimum reflectivity. The sensor's full width of half minimum/maximum (FWHM) is the difference of angles with 3 dBm output. The quality factor of the SPR sensor relates to the FWHM by the following: $\boldsymbol{Q} = \frac{\boldsymbol{\theta_\circ}}{\mathbf{FWHM}}$.

The figure of merit of an SPR biosensor is the sensor's sensitivity to the FWHM of the sensor's resonance dips ratio; $\frac{\mathbf{S}}{\mathbf{FWHM}}$. The Limit of detection is a vital parameter. It is the smallest amount of the pathogen the sensor can detect. Limit of detection can be calculated by $\frac{\mathbf{R}}{\mathbf{S_{surf}}}$. R is the resolution of the structure and $\mathbf{S_{surf}}$ is the ratio of the total refractive index change divided by the maximum surface of the mass concentration. $\frac{\Delta n}{\sigma_{max}}$. The uniformity, thickness and material of the thin metal layer can help improve the senor's performance. The refractive index of the prism, the output power of the light source and wavelength can be varied to optimize the sensor.

During this COVID-19 pandemic fight, accurate, cost-effective virus detection methods are vital. Optical sensors can provide accurate, rapid, cost-effective, rapid, and sensitive, technology for the COVID-19 detection. Improving and optimizing this sensors' structure to achieve the highest sensitivity and best limits of detection is the only way to make SPR biosensors a viable potential choice for detection of this pathogen; the COVID-19 virus.

4. Conclusion

The efforts to control the COVID-19 pandemic are vital to all. This paper gives a brief summary of this virus, the currently used detection technologies, the shortcomings of these technologies, and the potential of optical biosensing technologies development to for Covid-19 detection. Surface plasmon resonance biosensors are the primary focus of this summary due to SPR sensors' good sensitivity, resolution, response time, and limits of detection.

References

[1] Menni C, Valdes A M, Freidin M B, Sudre C H, Nguyen L H, Drew D A, Ganesh S, Varsavsky T, Cardoso M J, El-Sayed Moustafa J S et al., Jul 2020 "Real-time tracking of self-reported symptoms to predict potential covid-19," *Nat. Med.* **7** 1037-40.

[2] Li X, Zeng W, Li X, Chen H, Shi L, Li X, Xiang H, Cao Y, Chen H, Liu C et al., Apr 6 2020 "Ct imaging changes of corona virus disease 2019(covid-19): A multi-center study in southwest china," *J. Transl. Med.* **1** 154.

[3] Fang Y, Zhang H, Xie J, Lin M, Ying L, Pang P and Ji W, Aug 2020 "Sensitivity of chest ct for covid-19: Comparison to rt-pcr," *Radiology*. **2** E115-E17.

[4] Xie X, Zhong Z, Zhao W, Zheng C, Wang F and Liu J, Aug 2020 "Chest ct for typical coronavirus disease 2019 (covid-19) pneumonia: Relationship to negative rt-pcr testing," *Radiology*. **2** E41-E45.

[5] Samson R, Navale G R and Dharne M S, Sep 2020 "Biosensors: Frontiers in rapid detection of covid-19," *3 Biotech*. **9** 385.

[6] Nguyen T, Duong Bang D and Wolff A, Mar 14 2020 "2019 novel coronavirus disease (covid-19): Paving the road for rapid detection and point-of-care diagnostics," *Micromachines (Basel)*. **3** 306.

[7] Singh P, 2016 "Spr biosensors: Historical perspectives and current challenges," *Sens. Actuators B Chem.* 110-30.

[8] Uddin S M A, Chowdhury S S and Kabir E, May 25 2021 "Numerical analysis of a highly sensitive surface plasmon resonance sensor for sars-cov-2 detection," *Plasmonics*. 1-13.

[9] Huang L, Ding L, Zhou J, Chen S, Chen F, Zhao C, Xu J, Hu W, Ji J, Xu H et al., 2021/01/01/ 2021 "One-step rapid quantification of sars-cov-2 virus particles via low-cost nanoplasmonic sensors in generic microplate reader and point-of-care device," *Biosens. Bioelectron.* 112685.

[10] Kumar A, Yadav A K, Kushwaha A S and Srivastava S K, 2020 "A comparative study among ws2, mos2 and graphene based surface plasmon resonance (spr) sensor," *Sensors and Actuators Reports*. **1**

[11] Wang S, Shan X, Patel U, Huang X, Lu J, Li J and Tao N, Sep 14 2010 "Label-free imaging, detection, and mass measurement of single viruses by surface plasmon resonance," *Proc. Natl. Acad. Sci. U. S. A.* **37** 16028-32.

[12] Abid S A, Ahmed Muneer A, Al-Kadmy I M S, Sattar A A, Beshbishy A M, Batiha G E and Hetta H F, May 15 2021 "Biosensors as a future diagnostic approach for covid-19," *Life Sci.* 119117.

[13] Kaur M, Tiwari S and Jain R, Aug 2020 "Protein based biomarkers for non-invasive covid-19 detection," *Sens Biosensing Res.* 100362.

[14] Drobysh M, Ramanaviciene A, Viter R and Ramanavicius A, Apr 2 2021 "Affinity sensors for the diagnosis of covid-19," *Micromachines (Basel)*. **4** 390.

[15] Prabowo B A, Purwidyantri A and Liu K C, Aug 26 2018 "Surface plasmon resonance optical sensor: A review on light source technology," *Biosensors (Basel)*. **3** 80

Acknowledgments
This work was supported by the University of Malaya and INTI Research Grant Scheme 2020 [INTI-FITS-03-2020]

Optical Fiber Sensor for Heavy Chemical Detection: An Overview

Nurul Athirah Mohamad Abdul Ghafar[1], Arni Munira Markom[2*], Marni Azira Markom[3,4] and Ahmad Razif Muhammad[5].

[1] School of Electrical Engineering, College of Engineering, Universiti Teknologi MARA 40450 Shah Alam, Selangor, Malaysia.
[2] School of Electrical Engineering, College of Engineering, Universiti Teknologi MARA, Cawangan Johor, Kampus Pasir Gudang, 81750 Masai, Johor, Malaysia.
[3] Faculty of Electronic Engineering Technology, Universiti Malaysia Perlis, 02600 Arau, Perlis, Malaysia.
[4] Advance Sensor Technology, Centre of Excellence, Universiti Malaysia Perlis, Kampus Pauh Putra, 02600 Arau, Perlis, Malaysia.
[5] Institute of Microengineering and Nanoelectronics (IMEN), Universiti Kebangsaan Malaysia 43600 Bangi, Selangor, Malaysia.

[*]Corresponding author's email: arnimunira@uitm.edu.my

Abstract. Heavy metal contaminations such as mercury, lead, arsenic, cadmium, and zinc are becoming more serious and have become a hazard to human health. Due to their non-biodegradable nature, they can easily accumulate in the environment and cause toxicity even at low concentrations. Therefore, detecting the presence of these metal ions requires a highly sensitive sensing method. Traditional detection methods, such as electrochemical analysis, require complicated sample preparation, are costly, and typically require a lengthy measurement period. These days, optical fiber sensors have been acknowledged due to their unique characteristics such as compact size, high sensitivity, low cost, high flexibility, and immunity to electromagnetic interference. An overview of an optical fiber sensor technology for heavy chemical measurement is discussed in this paper. The sensing mechanisms are summarized, as well as the chemical water quality parameters and sensitivities.

1. Introduction

In recent years, the environmental degradation is crucial due to global development and enhancement of population, leads many researchers to investigate the negative impact as well as the solution such as for global warming, climate change, and species extinction [1]. One of the major factors is water source pollution. As we know water is vital source not only for human, but also for major species including animals and plants in the world. Most of developing countries like Malaysia is dependent on the conservative process of collecting water samples and water analysis [2-3]. Based on the river analysis in Malaysia, 150 out of 646 rivers in Peninsular Malaysia have been classified as polluted [4].

Many methods have been reported for water monitoring measurement and still, the most used due to accuracy and reliability is electrical conductivity devices [5]. However, this equipment can be significantly expensive, exhibits high power consumption and requires a professional person to handle the operation. Moreover, this technique is susceptible to electrical interference and does not consider other contents that affect water density but do not conduct electricity. Thus, a simple and cost-effective optical fiber sensor is chosen besides other advantages such as being immune to electromagnetic interference, compact size of design, high sensitivity, safe in hazardous environments and high durability against water corrosion [6]. This optical fiber is boosting a large researcher's attraction for wideband amplifiers, pulsed fiber laser from miro- to femtoseconds speed either in bright or dark pulses, and various types of fiber sensor development [7-11].

Thus, in this paper, we discussed various types of an optical fiber sensor for heavy chemical detection that purposely for river water monitoring. The sensitivity for each sensor is also elaborated with the specific heavy chemical such as Mercury (Hg), Nickel (Ni), Plumbum (Pb) and Ferum (Fe).

2. Heavy chemical effects to human's health

The most common heavy metals contaminants found in river water are mercury, lead, arsenic, chromium, cadmium, and zinc. These heavy chemicals are very dangerous to the human body and able to abolish the entire aquatic population. Mercury usually comes from instrument factories, coal burning, cosmetics waste, smelting process and salt electrolysis [12]. It can cause damage to the brain, kidneys and developing fetus. Thus, World Health Organization (WHO) has stated that there should only be 0.006 mg/L of mercury in drinking water [13].

Lead contaminants originate from various paints, cosmetics, hair dyes, gasoline, batteries, smelting, tableware, and coal. The maximum amount of lead that is safe to humans is 0.1 mg/L [13]. Excessive intake of lead can harm humans brain cells, cause mental retardation, brain death and dementia for elderly [14]. Activities such as mining, glass production, arsenic raw materials for pigments and papers can cause arsenic contamination if not properly handled.

WHO stated that the maximum amount of arsenic that can be consumed by humans is 0.01 mg/L at most [13]. Arsenic or arsenic compounds can lead to digestive and cardiovascular illnesses, harm to the nervous system, liver and skin cancer [15]. Chromium can cause numbness of limbs, ulcers, and mental disorders and thus only 0.05 mg/L are allowed in drinking water [13]. Chromium usually originated from cosmetics, leather preparations, tanned leather, chrome-plated metal parts, rubber, and ceramic raw materials [16]. Other contaminants that can be found in river water is cadmium.

It comes from waste activities in freshwater such as coal, mining, smelting and chemical industries and can cause renal dysfunction, cerebrovascular disease, high blood pressure and destroy the bone calcium. Thus, only 0.003 mg/L are allowed to be consumed by humans. Other than that, zinc also can be found amongst the contaminants. Zinc that mostly are from fertilizers, mining, steel processing and coal and waste combustion can cause stomach cramps, skin irritation, vomiting, nausea, anemia, causing respiratory disorder, and damage to the pancreas. It is advised not to take more than 5 mg/L [13].

3. Optical fiber sensor for heavy chemical detection

Table 1 shows the various optical sensor methods for different chemical measurements in water. Interferometer-based sensors have been commonly used for heavy metals detection. Z. Q. Tou et al. proposed an interferometer-based sensor using a photonic crystal fiber (PCF) that sandwiched between two single mode fibers (SMF) [17]. Impregnation and encapsulation technique were used to coat polyvinyl alcohol (PVA)/PAA hydrogel film on the PCF surface. 5-amino-8-hydroxyquinoline was modified chemically to allow Nikel ions (Ni^{2+}) been sensitive for detection. Then, a thermodynamically stable complex will be produced which decrease the volume gel and increasing the gel of refractive index on the surface of fiber, subsequently interference spectrum is shifted. The result from the experiment showed that the sensitivity of the sensor was 0.214 nm/uM.

Fiber grating utilizes the fiber materials photosensitivity in order to produce phase grating spatial in the core of fiber. The aim is to change periodically the refractive index of the core fiber in the specific grating area. J. Du et al. was successfully to fabricate a long period fiber grating (LPFG) with 25 mm long and 320 um period on normal SMF based on liquefied gas inscribed etch process [18]. Periodic mesoporous organosilica (PMO) films was grafted with selective ligand to perform as Lead ions (Pb^{2+}) sensitive films. This PMO will open and harmonize pore structure to allow metal material absorption.

Then, the film refractive index will be harmonizing matched closely to the fiber cladding refractive index, and thus lead concentration is sensitive to be detected. It is revealed that the sensitivity is about 2.072 nm/uM.

Table 1. Various types of fiber sensors and measurements.

Title & Citation	Method	Sensitivity	Chemical Measurement
Poly (vinyl alcohol) hydrogel based fiber interferometer sensor for heavy metal cations [17]	Interferometer based sensor coated with polyvinyl alcohol (PVA)/PAA hydrogel film	0.214 nm/μM	Ni^{2+}
Periodic Mesoporous Organosilica Films: Key Components of Fiber-Optic-Based Heavy-Metal Sensors [18]	Fiber grating-based sensors grafted with Periodic mesoporous organosilica (PMO) films	2.072 nm/μM	Pb^{2+}
Surface Plasmon Resonance based fiber optic sensor for mercury detection using gold nanoparticles PVA hybrid [19]	Localized Surface plasmon resonance (LSPR) optical fiber sensor coated with gold nanoparticles-polyvinyl alcohol (PVA) complex	-	Hg^{2+}
Optical fiber-based heavy metal detection using the Localized Surface Plasmon Resonance technique [20]	Localized Surface plasmon resonance (LSPR) optical fiber sensor coated with gold nanoparticles	0.28 nm/mM	Pb^{2+}
A Fiber-optic Sensor Based on Plasmon Coupling Effects in Gold Nanoparticles Core-satellites Nanostructure for Determination of Mercury Ions (II) [21]	Localized Surface plasmon resonance (LSPR) optical fiber sensor coated with gold nanoparticles	-	Hg^{2+}
Chitosan/poly(acrylic acid) based fiber-optic surface plasmon resonance sensor for Cu^{2+} ions detection [22]	Surface plasmon resonance (LSPR) optical fiber sensor using Tapering method coated with gold film and chitosan	0.1184 nm/μM	Cu^{2+}
Ratiometric fluorescence sensor for Fe^{3+} ions detection based on quantum dot-doped hydrogel optical fiber [23]	Ratiometric fluorescence sensor coated with thioglycolic acid and graphene quantum dots	-	Fe^{3+}

Surface plasmonic is a technique occurred between metal and dielectric interface that widely used for monitoring the analytes testing in the biochemical measurements. The plasmonic fiber sensors have many advantages such as high integration capability, small volume, and real-time remote monitoring and detection compared to the conventional method of prism-based plasmonic sensors. D. Raj et al. presented a surface plasmon resonance (SPR) fiber sensor with gold nanoparticles-polyvinyl alcohol (PVA) complex to enhance the sensitivity of Mercury ions (Hg^{2+}) [19]. A plastic-coated quartz fiber with 630 um fiber diameter and 600 um core diameter is used for the sensor development. Figure 1 illustrated the schematic diagram of experimental configuration. A sensitive surface area where the cladding is removed at about 3cm was coated by thermal evaporation. The SPR spectrum will be shifted and induced when gold embedded in PVA is contacted to Hg^{2+} ions. The result from the experiment showed the detection limit was 1 μM but no sensitivity result is stated.

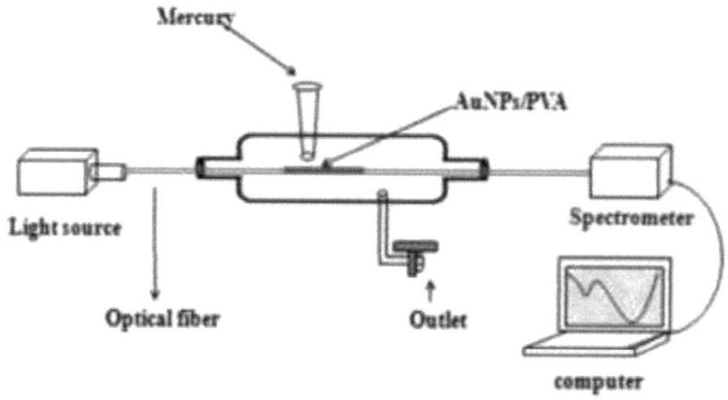

Figure 1. Schematic diagram of experimental setup. Reproduced from Ref. [19].

A reflection-based localized surface plasmon resonance (LSPR) fiber-optic sensor was demonstrated by P. Dhara et al. in 2019 to detect Pb^{2+} ions [20]. Gold nanoparticles have been immobilized using self-assembling method on the surface of optical fiber with 1,1- Mercaptoundecanoic acid to increase the sensitivity towards Pb^{2+} ions. The heavy metal concentration can be determined using the correlation between LSPR resonance shift wavelength and binding rate. Ethylenedinitrilotetraacetic acid was used to remove the bound of heavy metal ions and the Pb^{2+} ion sensitivity of the sensor was 0.28 nm/mM.

S. Jia et al. proposed using the same method of LSPR fiber sensor for Hg^{2+} ions measurement [21]. This method was based on the gold nanoparticles plasmon coupling effect that linked to a specific Mercury ions and thymine (T)-rich DNA hybridization. The fiber sensing probe is fabricated using a multimode fiber (MMF) with 600 um core diameter of fiber. Electrostatic self-assembly process was used to coat gold nanoparticles on the sensitive area of fiber to allow LSPR excitement. A reflective silver film was produced at the end of the fiber by the silver mirror reaction and then, it will be reflecting the signal light to enable sensing reflection. The result from the experiment showed the detection limit was 3.4 nM whereas the sensitivity was not reported.

A surface plasmon resonance fiber optic based on tapering method for Copper ions (Cu^{2+}) measurement was successfully presented by Ding et al. in 2019 [22]. Two identical MMFs were used with a length of 1.6 cm tapering no-core fiber spliced in between of the fibers. A 1.5 cm long of the fiber coreless was spliced in between of the two MMFs using an electric arc splicer before it was heated by a tapering flame equipment to be stretched up to 1.6 cm. A gold film and chitosan were fabricated to coat the outer face of the no-core fiber tapered. The gold films will stimulate the surface plasmon waves and creating the sensitivity of sensor for Cu^{2+} ions measurement. It was observed that low concentrations will give the best sensor performance whereas the increasing concentration cause the decreasing sensitivity. The Cu^{2+} ions concentration was ranging between 0.2 to 50 um to achieve 0.1184 nm/μM sensitivity.

Finally, fluorescence effect sensing concept for heavy metal detection was produced from operationalized molecules on the fiber sensor, in order to develop non-covalent bonding or changes in chemical reaction, and thus photophysical properties change of the fluorescent chromophore was induced. Then, the fluorescence signal changes will be recognized and detect the heavy metal ions. M. Zhou et al. was demonstrated the Cadmium Telluride (CdTe) doped with hydrogel for Ferric ions (Fe^{3+}) detection [23]. A green emissive thioglycolic acid capped graphene quantum dots (gQDs) coated the fiber sensitive area to quench selectively Fe^{3+} ions with a red emissive N-Acetyl-lcysteine capped QDs (rQDs) that immunized against Fe^{3+} ions. By measuring the ratio of fluorescence intensities of two QDs, the detection limit was 14 nM but again no sensitivity result was reported.

Conclusion

In this paper, the current research of heavy metal ion sensors based on optical fiber technology was briefly reviewed. Various sensing methods such as Interferometer based sensor, Fiber grating-based sensors, Localized Surface plasmon resonance (LSPR) and Ratiometric fluorescence sensor were presented. Thus, the surface plasmon resonance concept with tapering method achieves the best sensitivity for chemical detection with 0.1184 nm/μM measurement.

Water quality monitoring are essential especially freshwater for the life of human, animals and plants in this world. Therefore, there are many papers on heavy metal ion sensors for monitoring were published, however their sensitivity results of sensor are often neglected and not included. In fact, the sensitivity of the sensors is important to know the sensor's performance and accuracy. It also will allow other researchers to study and investigate these sensors to make an improvement for the future sensing development.

References

[1] Irigoyen, M., Sánchez-Martin, J. A., Bernabeu, E., & Zamora, A. Tapered optical fiber sensor for chemical pollutants detection in seawater. 2017 *Measurement Science and Technology*, **28**(4), 045802.

[2] Razali, A., Ismail, S. N. S., Awang, S., Praveena, S. M., & Abidin, E. Z. Land use change in highland area and its impact on river water quality: a review of case studies in Malaysia. 2018 *Ecological processes*, **7**(1), 1-17.

[3] Goi, C. L. The river water quality before and during the Movement Control Order (MCO) in Malaysia. 2020 *Case Studies in Chemical and Environmental Engineering*, **2**, 100027.

[4] Liang, Y. Q., Annammala, K. V., Martin, P., Yong, E. L., Mazilamani, L. S., & Najib, M. Z. M. Assessment of Physical-Chemical Water Quality Characteristics and Heavy Metals Content of Lower Johor River, Malaysia. 2020 *Journal of Environmental Treatment Techniques*, **8**(3), 961-966.

[5] Rahman, H. A., Harun, S. W., Yasin, M., Phang, S. W., Damanhuri, S. S. A., Arof, H., & Ahmad, H. Tapered plastic multimode fiber sensor for salinity detection. 2011 *Sensors and Actuators A: Physical*, **171**(2), 219-222.

[6] Zain, H. A., Jali, M. H., Rahim, H. R. A., Johari, M. A. M., Yusof, H. H. M., Thokchom, S., ... & Harun, S. W. ZnO nanorods coated microfiber loop resonator for relative humidity sensing. 2020 *Optical Fiber Technology*, **54**, 102080.

[7] Markom, A. M., Paul, M. C., Dhar, A., Das, S., Pal, M., Bhadra, S. K., ... & Harun, S. W. Performance comparison of enhanced Erbium–Zirconia–Yttria–Aluminum co-doped conventional erbium-doped fiber amplifiers. 2017 *Optik*, **132**, 75-79.

[8] Chong, A., Buckley, J., Renninger, W., & Wise, F. All-normal-dispersion femtosecond fiber laser. 2006 *Optics express*, **14**(21), 10095-10100.

[9] Markom, A. M., Tan, S. J., Haris, H., Paul, M. C., Dhar, A., Das, S., & Harun, S. W. Experimental observation of bright and dark solitons mode-locked with zirconia-based erbium-doped fiber

laser. 2018 *Chinese Physics Letters*, **35**(2), 024203.

[10] Markom, A. M., Tan, S. J., Muhammad, A. R., Paul, M. C., Dhar, A., Das, S., ... & Harun, S. W. Dark pulse mode-locked fibre laser with zirconia-based erbium-doped fibre (Zr-EDF) and Black phosphorus saturable absorber. 2020 *Optik*, **223**, 165635.

[11] Qian, Y., Zhao, Y., Wu, Q. L., & Yang, Y. Review of salinity measurement technology based on optical fiber sensor. 2018 *Sensors and Actuators B: Chemical,* **260**, 86-105.

[12] Chen, C.-W., Chen, C.-F., & Dong, C.-D. Distribution and Accumulation of Mercury in Sediments of Kaohsiung River Mouth, Taiwan. 2012 *APCBEE Procedia*, **1**, 153–158.

[13] World Health Organization. Guidelines for drinking-water quality: World Health Organization. 2011 *Distribution and Sales, Geneva*, *27*.

[14] Martin, S., & Griswold, W. Human health effects of heavy metals. 2009 *Environmental Science and Technology briefs for citizens*, **15**, 1-6.

[15] Singh, N., Kumar, D., & Sahu, A. P. Arsenic in the environment: effects on human health and possible prevention. 2007 *Journal of environmental biology*, **28**(2), 359.

[16] Ghani, A., & Ghani, A. Effect of chromium toxicity on growth, chlorophyll and some mineral nutrients of brassica juncea L. 2011 *Egyptian Academic Journal of Biological Sciences, H. Botany*, **2**(1), 9-15.

[17] Tou, Z. Q., Koh, T. W., & Chan, C. C. Poly (vinyl alcohol) hydrogel based fiber interferometer sensor for heavy metal cations. 2014 *Sensors and Actuators B: Chemical*, **202**, 185-193.

[18] Du, J., Cipot-Wechsler, J., Lobez, J. M., Loock, H. P., & Crudden, C. M. Periodic Mesoporous Organosilica Films: Key Components of Fiber-Optic-Based Heavy-Metal Sensors. 2010 *Small*, **6**(11), 1168-1172.

[19] Raj, D. R., Prasanth, S., Vineeshkumar, T. V., & Sudarsanakumar, C. Surface Plasmon Resonance based fiber optic sensor for mercury detection using gold nanoparticles PVA hybrid. 2016 *Optics Communications*, **367**, 102-107.

[20] Dhara, P., Kumar, R., Binetti, L., Nguyen, H. T., Alwis, L. S., Sun, T., & Grattan, K. T. Optical fiber-based heavy metal detection using the localized surface plasmon resonance technique. 2019 *IEEE Sensors Journal*, **19**(19), 8720-8726.

[21] Shuo, J., Chao, B., Jian-Hua, T. O. N. G., Ji-Zhou, S., & Shan-Hong, X. A Fiber-optic sensor based on plasmon coupling effects in gold nanoparticles core-satellites nanostructure for determination of mercury ions (II). 2017 *Chinese Journal of Analytical Chemistry*, **45**(6), 785-790.

[22] Ding, Z. W., Ravikumar, R., Zhao, C. L., Chen, L. H., & Chan, C. C. Chitosan/Poly (acrylic acid) based fiber-optic surface plasmon resonance sensor for Cu 2+ ions detection. 2019 *Journal of Lightwave Technology*, **37**(10), 2246-2252.

[23] Zhou, M., Guo, J., & Yang, C. Ratiometric fluorescence sensor for Fe3+ ions detection based on quantum dot-doped hydrogel optical fiber. 2018 *Sensors and Actuators B: Chemical*, **264**, 52-58.

Acknowledgement
This work is sponsored by Optical Society Malaysia (OSM) and Ministry of Education (MoE) Malaysia under grant 600-IRMI/FRGS-RACER 5/3 (107/2019).

Characterization of cost-efficient OPT101 PIN photodiode as relative dosimetry in diagnostic radiology: An IOT application

Y Md Radzi[1*], M K Zainol[1], and A S A Al Naggar[1]

[1] School of Physics, Universiti Sains Malaysia, 11800, Penang, Malaysia

[*]Corresponding author's email: yasminradzi@usm.my

Abstract. OPT101 photodiode sensor dosimetry properties were investigated for various parameters to observe its capability as a relative dosimeter. Its characterization was done with PTW semiconductor detector for diagnostic radiology as a benchmark. OPT101 presented excellent output linearity relative to dose measured by PTW semiconductor for energy dependency and reproducibility test. A significant high sensitivity was observed against dose rates and mean energies and linear response against different source-to-detector distance, SDD. Results suggest the use of an Arduino IDE microcontroller as potential of utilizing an IoT application in input/output communication with the sensor and data processing.

Index Terms. OPT101 photodiode, relative dosimeter, Arduino IDE

1. Introduction

An ideal dosimeter must possess at least one of the characteristics for it to function as a radiation dosimeter, which are linearity, precision and accuracy, dose rate dependence, energy and directional dependency [1-2]. A photodiode is the alternative to the silicon diode dosimeter because of its has small size, real-time operation, low cost, high accuracy, and precision attainable in a wide range of doses [3-4]. The only lack of photodiode is its sensitivity that varied with temperature, dose rate, and its efficiency degraded over time. It is increasingly used for small field dosimetry in modern therapy techniques due to their finite size of the sensitive volume able to produce 18, 000 times current than that of an ionization chamber of equal sensitive volume [5]. In this work, the immediate read-out signal from the photodiode is facilitated by the used of IoT system applications from Arduino IDE software and Blynk app available in major mobile operating systems nowadays.

2. Method

OPT101 monolithic photodiode was connected onto the Arduino Uno board with LED connected within the circuit to ensure a closed loop is established. The whole transducer system is then exposed under x-ray irradiation alongside with PTW semiconductor detector (Freiburg, GmbH) for diagnostic x-ray. The photovoltaic signals were extracted (using a digital multimeter) under a set of exposure settings, including tube voltage (kVp), current-time product (mAs) and source-to-detector distance (SDD). Transducer performance analysis was based on signal linearity to mAs and air kerma, and sensitivity dependence on absorbed dose, energy, and dose rate.

On the other hand, Arduino code is written and compiled using Arduino interfacing software (Sketch) to collect the signal generated by the transducer and send it to the Arduino Uno board via USB cable. The whole system transducer with Arduino is optionally integrated with Blynk application via Wi-Fi, GSM, or Ethernet cables to monitor the dose response of the OPT101 photodiode on user's mobile phone.

3. Results and Discussion

OPT101 photodiode dependency against kilovoltage peak (kVp) and dependency against current-time product (mAs), both indicated a linear behavior, with R^2 values of 0.964 and 0.999 respectively. Fig. 1. (a) OPT101 photodiode sensitivity dropped from 15.32 mV/mGy (at a dose rate of 0.83 mGy/s) to 3.74 mV/mGy (at a dose rate of 1.66 mGy/s) before becoming stable beyond dose rate of 2 mGy/s. While Fig.1. (b) OPT101 sensitivity continuously fading from 2.65 mV/mGy (at tube voltage of 50 kVp) to 1.11 mV/mGy (at tube voltage of 70 kVp) and become stable beyond approximately 75 kVp tube voltage.

Fig.2. illustrated that OPT101 signal linearity to dose determinant coefficients were 0.946 with a small signal detected per absorbed dose increment (0.6 mV/mGy). On the other hand, OPT101 photodiode sensitivity dropped from 15.3 mV/mGy to 2.5 mV/mGy in response within a 20.7 mGy absorbed dose increment similar to what has been observed previously [6-8].

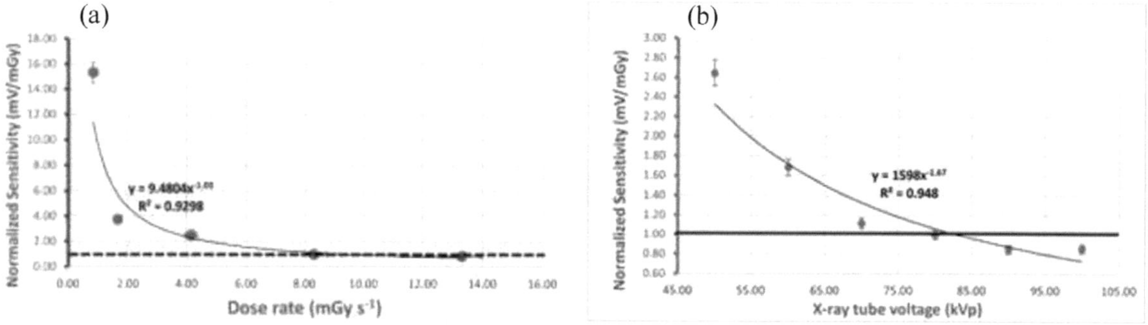

Fig. 1. Sensitivity of OPT101 photodiode across different (a) dose rates (range 10 mA – 160 mA) and (b) X-ray tube voltage (at 100 mAs); both at 100cm SDD.

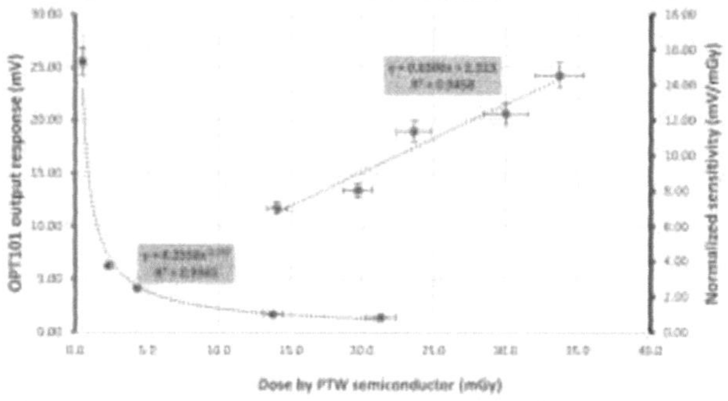

Fig. 2. Output signal and transducer sensitivity as a function of dose (measured by PTW semiconductor dosimeter). This data was collected in the 10–160 mAs range, while the tube voltage and SDD were set to 80 kVp and 100 cm, respectively.

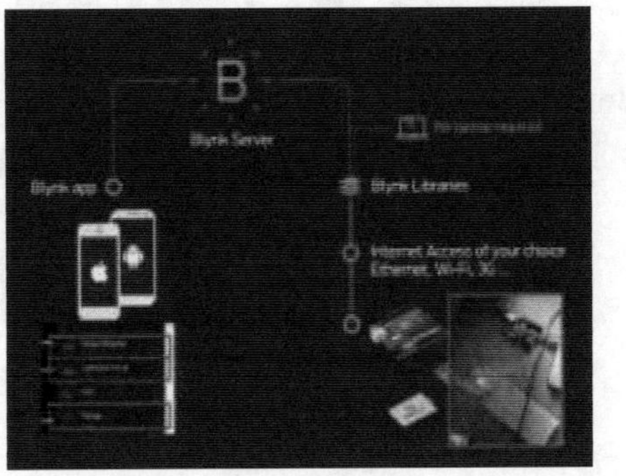

Fig. 3. Architecture diagram of integration of Arduino IDE with the Blynk IoT application [9].

The results highlight two important parameters: (i) OPT101 photodiode has a potential to be optimized as a relative dosimeter in the future application of diagnostic x-ray detector with stated characteristic mentioned above. However, it has experienced a low radiation-induced signal/sensitivity possible due to less charge drift in increasing energy delivered to photodiode active area. Transducer sensitivity and response to X-ray beams is directly influenced by the energy required to form an Electron-Hole Pair in the transducer's active region. Hence, this observation agrees with a previous study where photo transistor sensitivity decreased in response to increasing dose-rate [10]. Furthermore, considering the interaction events take place in lower x-ray energy (predominantly Photoelectric effect and Compton scattering) within the active area of photodiode, OPT101 photodiode having advantages of high sensitivity with immediate readout of dose values, low cost, and portability also could replace passive dosimetry such as thermoluminescent dosimeter (TLD) for dose measurements in diagnostic radiology.

Secondly, (ii) the whole transducer system can be well-versed into a sophisticated IoT integration with less set-up required (no laptop and messy cords, only Wi-Fi and Blynk server) for the future data collection as illustrated in Fig. 3. Arduino IDE can easily be connected to the cloud and build a no-code iOS, Android, and web applications to read and analyze real-time data coming from OPT101 sensor, control them remotely from anywhere (e.g. mobile phone) and receive important notifications.

References

[1] Seco, J., Clasie, B., & Partridge, M. (2014). Review on the characteristics of radiation detectors for dosimetry and imaging. Physics in Medicine and Biology, 59(20), R303–R347.

[2] Izewska J., Rajan G., 2005. "Radiation dosimeters," Document IAEA 08/2005, STI/PUB/1196, Textbook, Chapter3, 71–99.

[3] Romei, C., Di Fulvio, A., Traino, C. A., Ciolini, R., & d'Errico, F. (2015). Characterization of a low-cost PIN photodiode for dosimetry in diagnostic radiology. Physica Medica, 31(1), 112–116.

[4] Oliveira, C. N. P., Khoury, H. J., & Santos, E. J. P. (2016). PiN photodiode performance comparison for dosimetry in radiology applications. Physica Medica, 32(12), 1495–1501. H. Zhu et al., "Low-threshold GaN thin-film random laser through the weak scattering feedback," J. Phys. D. Appl. Phys., vol. 50, no. 4, p. 045107, Feb. 2017.

[5] Yarahmadi M, Wegener S, Sauer OA. Energy and field size dependence of a silicon diode designed for small-field dosimetry. Med Phys. 2017 May;44(5):1958-1964. doi: 10.1002/mp.12195. Epub 2017 Apr 12. PMID: 28273357.

[6] Damulira, E., Yusoff, M. N. S., Omar, A. F., & Taib, N. H. M. (2019). A review: Photonic devices used for dosimetry in medical radiation. Sensors (Switzerland), 19(10), 1–28.

[7] Nazififard, M., Suh, K. Y., & Mahmoudieh, A. (2016). Experimental analysis of a novel and low-cost pin photodiode dosimetry system for diagnostic radiology. Review of Scientific Instruments, 87, 073502

[8] Md Radzi Y., Zulkafli N., Omar A. Preliminary investigation on performance of photodiode

sensor as a dosimeter. Biomedical Photonics. 2021; 10 (1): 4-10.
[9] Blynk Inc. (2021). Introduction to Blynk. Blynk Documentation. https://docs.blynk.io/en/
[10] Andjelkovi´c, M.S., Risti´c, G.S., 2015. Current mode response of phototransistors to gamma radiation. Radiat. Meas. 75, 29–38

Design of Multiwavelength EDF-Raman Lasers Utilizing Mach-Zehnder Interferometer

Nani Fadzlina Naim[1*], Awangku Nur Azree Awang Azlan[1], Muhammad Faiz Ibrahim[2], Suzi Seroja Sarnin[1], Norsuzila Ya'acob[1], Mohd Saiful Dzulkefly Zan[3]

[1] Faculty of Electrical Engineering, School of Engineering, Universiti Teknologi MARA, 40450 Shah Alam, Selangor, Malaysia.
[2] TNB Research SDN. BHD., 43000 Kajang, Selangor, Malaysia.
[3] Department of Electrical, Electronic and Systems Engineering, Faculty of Engineering and Built Environment, Universiti Kebangsaan Malaysia, Bangi 43600, Malaysia

[*]Corresponding author's email: nanifadzlina@uitm.edu.my

Abstract. This paper presents the design of multiwavelength Erbium Doped Fiber (EDF)-Raman fiber laser utilizing Mach-Zehnder Interferometer (MZI) with various cavity structures. A multiwavelength laser employing hybrid gain medium of EDF-Raman amplifier is simulated using OptiSystem software. Three cavity structures of multiwavelength laser such as unidirectional, bidirectional and ring cavity are simulated and analysed. From the simulation result, it is found that ring cavity structure produced the best performance whereas at 1000 mW pump power, up to 19 lasing lines were obtained in the ring cavity, compared to 18, and 16 lasing lines in the unidirectional and bidirectional linear cavity, respectively. All multiwavelength fiber lasers exhibit the same line spacing of 4.9 nm. In addition, at coupling coefficient of 0.9, up to 49.5 dB of side mode suppression ratio (SMSR) were achieved in the ring cavity structure, compared to 49.2 dB, and 30.3 dB of SMSR in the unidirectional and bidirectional linear cavity, respectively. However, bidirectional linear cavity exhibits the highest peak power of 2.07 dBm, compared to -17.4 dBm, and -15.5 dBm of peak power in the unidirectional linear and ring cavity, respectively.

1. Introduction

Multiwavelength fiber laser sources are very attractive and have potential applications in instrument testing, sensing, temperature sensors [1], optical fiber sensors as well as dense wavelength-division-multiplexing (WDM) communication systems [2]. In designing a multiwavelength fiber laser, it is important to know how to design of multiwavelength fiber laser as well as its specifications such as pump power, optical fiber length, polarization state, coupling ratio, and so on. There are optical fibers that perform as a gain medium such as erbium doped fiber [3-5] or Raman fiber [6][7] in multiwavelength fiber laser. Other than that, semiconductor optical amplifier as a gain medium also has proposed in multiwavelength fiber laser [8][9]. In the development of optical fiber communications that can operate over long range, erbium doped fiber amplifiers (EDFA) seem to be the frequently used fiber amplifiers. The two wavelengths; 980 nm and 1480 nm are commonly used to pump an EDFA and achieve gain around 1550 nm for C-Band and 1590 nm for L-Band. There are many multiwavelength fiber lasers have been developed that utilize erbium doped fiber (EDF) technology to generate multiwavelength laser operation. Ahmed et al. [10] recently reported a multiwavelength laser source in a linear cavity structure, in which EDF technology is used in the cavity to generate a stable multiwavelength laser operation. However, due to the strong homogenous broadening effect, generating a stable multiwavelength output with erbium doped fiber is challenging. Recently, multiwavelength fiber laser that utilized a hybrid gain medium [11] can overcome the strong homogenous broadening effect. Wang et al. [12] successfully accomplished stable multiwavelength laser operation with a multiwavelength fiber laser in a ring cavity structure that integrated SOA-EDF technology to form a hybrid gain medium. Apart from EDF technology, multiwavelength fiber laser using a raman amplifier as gain medium have been reported and proposed frequently for a couple of years due to their significant advantages. Raman amplifiers are often more expensive and have lower gain than EDF. As a result, it is indeed employed in a handful of applications. Ultra-wideband amplifiers, low noise figure, as well as a broad range of applications are some of the features of the Raman amplifier. This advantage enables optical fiber communication utilizing a Raman Amplifier to transmit data over vast distances.

In this paper, multiwavelength EDF-Raman fiber laser with various cavity structures, such as unidirectional linear, bidirectional linear, and ring cavity structures, will be investigated and simulated. Aside from that, the optimum EDF and EDF-Raman gain was obtained by varying the EDF and Raman Amplifier lengths. The ring cavity structure was found to be the best structure when compared to unidirectional and bidirectional linear cavity structures. For the ring cavity structure, up to 19 lasing lines with SMSR of 49.5 dB are achieved.

2. Result and Discussion

The multiwavelength fiber laser is designed and simulated using OptiSystem software. Fig. 1 (a), 1 (b), and 1 (c) show the schematic diagram of a multiwavelength fiber laser for unidirectional linear, bidirectional linear, as well as ring cavity structures, respectively. The structures in Fig. 1 (a), and 1 (b) resemble a linear cavity, whereas in Fig. 1 (c) resembles a ring cavity. The gain medium in the laser cavity is the EDF with its length is set to 10 m, as well as the Raman Amplifier with its length is set to 12 km. For the EDF, a laser diode, LD1 with a wavelength of 1480 nm and a pump power of 50 mW is used to generate the EDF amplification. The pump power for both LD2 and LD3 is 50 mW, respectively. The ideal multiplexer will multiplex the signals from the two laser diodes; LD2 and LD3 with wavelengths of 1450 nm and 1460 nm, respectively.

Table 1 shows the side mode suppression ratio (SMSR), peak power and number of lasing lines obtained when different coupling ratio is applied were successfully obtained for all the three structures. The coupling ratios are varied from 0.25, 0.50, 0.75 and 0.9. When using a 0.9 coupling ratio, it was found that ring cavity structure had the optimum performance in terms of SMSR of 49.5 dB. In addition, unidirectional linear and bidirectional linear cavity structures exhibit SMSR of 49.2 dB and 30.3 dB, respectively, when 0.9 coupling ratio is used. In addition, ring cavity has the most lasing lines with a total of 19 lasing lines as compared to unidirectional and bidirectional cavity.

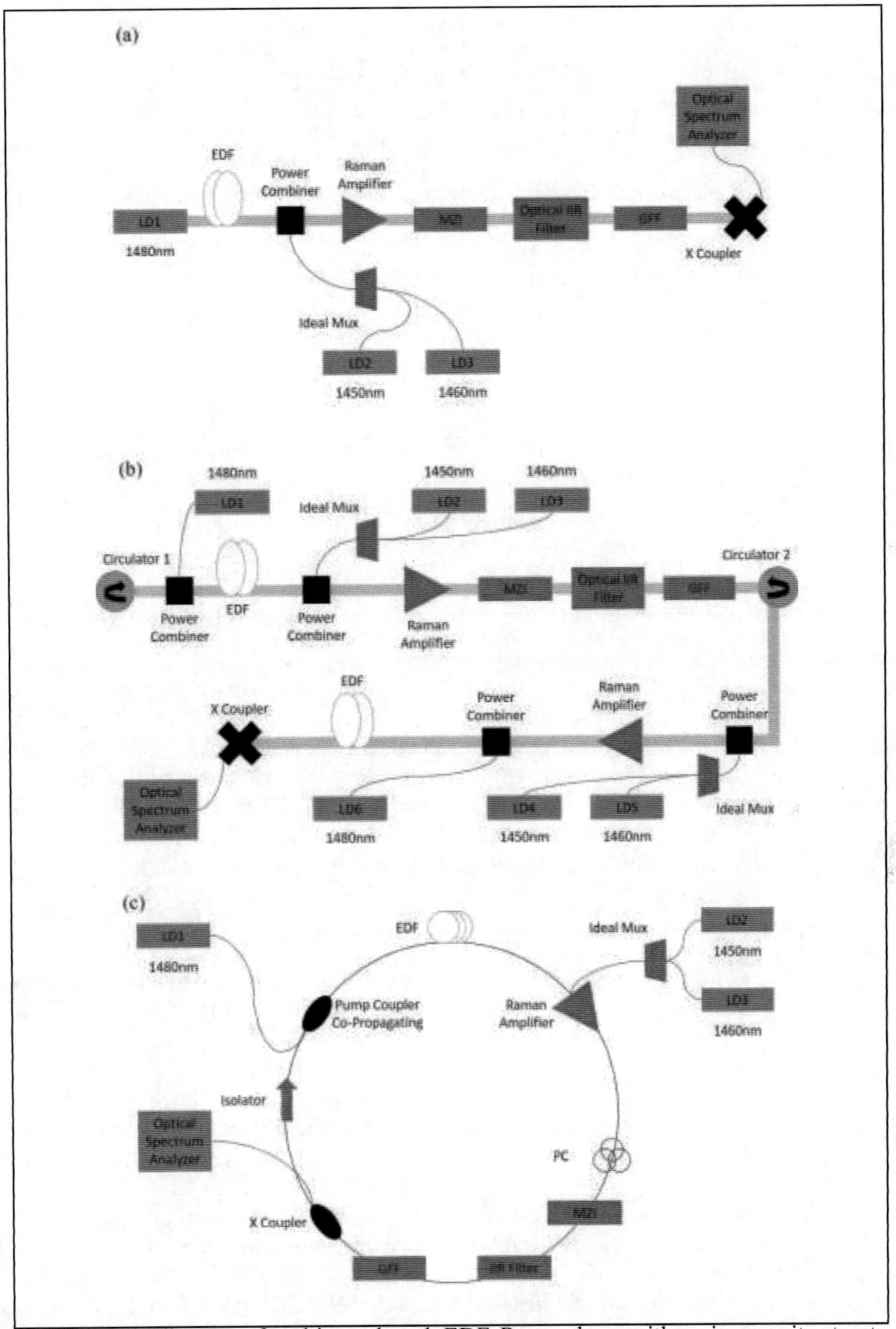

Figure 1. Schematic diagram of multiwavelength EDF-Raman laser with various cavity structures. (a) Unidirectional linear cavity structure. (b) Bidirectional linear cavity structure. (c) Ring cavity structure. EDF is Erbium Doped Fiber, MZI is Mach-Zehnder Interferometer, LD is laser diode and GFF is Gain Flattening Filter,

Table 1. Number of lasing lines, peak power and average SMSR with different coupling ratio for various Multiwavelength laser structures

Structure	Coupling Ratio	Number of lasing lines	Peak Power (dBm)	Average SMSR (dB)
Unidirectional Linear	0.25	17	-23.4	48.0
	0.50	18	-20.4	48.3
	0.75	18	-18.7	48.9
	0.90	18	-17.9	49.2
Bidirectional Linear	0.25	15	-3.9	31.5
	0.50	15	-1.1	31.5
	0.75	16	0.7	30.3
	0.90	17	1.6	30.3
Ring	0.25	18	-24.6	47.8
	0.50	18	-21.6	48.8
	0.75	18	-19.9	49.3
	0.90	19	-19.0	49.5

3. Conclusion

In conclusion, multiwavelength EDF-Raman fiber laser with various structures were successfully designed and simulated using OptiSystem software. By varying the EDF and Raman Amplifier lengths, the optimum EDF and EDF-Raman gain was obtained when EDF and Raman Amplifier length are 10 m and 12 km, respectively. In comparison to the other structures, it was found that the ring cavity structure has the greatest number of lasing lines and the highest SMSRs with 19 lasing lines and 49.5 dB of SMSRs is the preferred design. Furthermore, the number of lasing lines, peak power, and SMSR can be flexibly obtained by varying the EDF length, Raman Amplifier length and pump power.

References

[1] Chen, B., Wang, T., Jia, Q., Wang, Z., & Zhao, D. (2018). High sensitive temperature sensing system based on multi-wavelength fiber laser with dispersion shifted fiber. OSA Continuum, 1(2), 401–407.

[2] Qureshi, K. K., Tam, H. Y., Chung, W. H., & Wai, P. K. A. (2005). Multiwavelength laser source using linear optical amplifier. IEEE Photonics Technology Letters, 17(8), 1611–1613.

[3] Azhan, M. N. M., Cholan, N. A., Talib, R., & Sulaiman, A. H. (2018). Multiwavelength Brillouin-erbium Fiber Laser in Linear Cavity. 2018 IEEE 7th International Conference on Photonics (ICP), 1–3.

[4] He, W., Li, D., Zhu, L., Dong, M., & Luo, F. (2017). Tunable Multiwavelength Erbium-Doped Fiber Laser Employing PM-FBG and Mach–Zehnder Interferometer with Optical Fiber Delay Line. IEEE Photonics Journal, 9(3), 1–8.

[5] Al-Mansoori, M., Abdullah, M. K., Mahdi, M. A., & Iqbal, S. J. (2005). Threshold Features of L-Band Linear Cavity Multiwavelength Brillouin-Erbium Fiber Laser. TENCON 2005 - 2005 IEEE Region 10 Conference, 1–4.

[6] Qin, S., Chen, D., Tang, Y., & He, S. (2006). Stable and uniform multi-wavelength fiber laser based on hybrid Raman and Erbium-doped fiber gains. Optics Express, 14(22), 10522–10527.

[7] Abass, A. K., Al-Mansoori, M. H., Jamaludin, M. Z., Abdullah, F., Al-Mashhadani, T. F., & Ali, M. H. (2013). Multiwavelength Brillouin-Raman fiber laser utilizing linear and ring cavity

structures: A comparative study. 2013 IEEE 4th International Conference on Photonics (ICP), 138–140.

[8] Jian Yao, Jianping Yao, Zhichao Deng, & Jian Liu. (2005). Multiwavelength erbium-doped fiber ring laser incorporating an SOA-based phase Modulator. IEEE Photonics Technology Letters, 17(4), 756–758.

[9] O'Riordan, C., Connelly, M. J., Evans, I., Anandarajah, P. M., Maher, R., & Barry, L. P. (2007). Actively Mode-locked Multiwavelength Fibre Ring Laser Incorporating a Lyot Filter, Hybrid Gain Medium and Birefringence Compensated LiNbO3 Modulator. 2007 9th International Conference on Transparent Optical Networks, 248–251.

[10] Ahmed, F., Kishi, N., & Miki, T. (2005). Multiwavelength erbium-doped fiber Fabry-Pe/spl acute/rot laser and its uniform spectral lines power operation. IEEE Photonics Technology Letters, 17(4), 753–755.

[11] Yumeng Tu, Huaping Gong, Leilei Zhang, Yongxing Jin, & Jiangfeng Wang. (2011). Multiwavelength Brillouin fiber laser with EDFA. Proceedings of 2011 International Conference on Electronics and Optoelectronics, 186–188.

[12] Wang, D. N., Tong, F. W., Fang, X., Jin, W., Wai, P. K. A., & Gong, J. M. (2003). Multiwavelength erbium-doped fiber ring laser source with a hybrid gain medium. Optics Communications, 228(4–6), 295–301.

Acknowledgment
The author wishes to thank Universiti Teknologi MARA (UiTM) under internal research grant 600-RMC/GPK 5/3 (101/2020) for supporting this project.

The generation of nanosecond pulses at C-band region with titanium dioxide as a saturable absorber

N Ahmed[1*], S Omar[1], NF Zulkipli[5], Z Jusoh[2], HA Rahman[4], B Musa[3] and SW Harun[5,6]

[1]School of Electrical Engineering, College of Engineering, University Teknologi Mara Cawangan Terengganu Kampus Dungun, 23000 Dungun, Malaysia
[2]School of Electrical Engineering, College of Engineering, University Teknologi Mara Cawangan Terengganu Kampus Kuala Terengganu, 21080 Chendering, Malaysia
[3]School of Electrical Engineering, College of Engineering, University Teknologi Mara Cawangan Terengganu Kampus Bukit Besi, 23000 Dungun, Malaysia
[4]School of Electrical Engineering, College of Engineering, University Teknologi Mara, 40450 Shah Alam, Malaysia
[5]Department of Physics, Faculty of Science and Technology, Airlangga University, Surabaya, Indonesia
[6]Photonics Engineering Laboratory, Department of Electrical Engineering, Faculty of Engineering, University of Malaya, 50603 Kuala Lumpur, Malaysia

[*]Corresponding author's email: noriz207@uitm.edu.my

Abstract. A passive mode-locked in an erbium-doped fiber laser with titanium dioxide (TiO_2) film as a saturable absorber (SA) has been successfully demonstrated. The film is fabricated using a liquid phase exfoliation, which offers a simple and low-cost method. The self-starting mode-locked was created by inserting a 200-meter-long single-mode fiber into the laser cavity to balance the nonlinearity and dispersion of the cavity. The pulses operate stably at a central wavelength of 1560 nm. The pulse repetition rate was almost fixed at 988 kHz at a tuneable pump power from 145.83 mW to 187.04 mW. The repetition rate shows excellent stability with a signal-to-noise ratio (SNR) of 69 dB whilst the pulse width was virtually constant at 230 ns. The maximum output power was measured at 2.17 mW, eliciting maximum pulse energy of 2.19 nJ. This experiment demonstrates that stable mode-locked pulsed can be generated using TiO_2-SA.

1. Introduction

The pulses with a nanosecond-duration generated by mode-locked fiber lasers have attracted broad attention in many applications in a variety of fields, including remote sensing, welding industries, and medical [1–3]. Beforehand, electro-optic and acousto-optic modulators were used to actively generate nanosecond pulses. However, this active modulation approach necessitates a large and costly component. Therefore, passive techniques are preferable to active techniques for producing mode-locked EDFLs because they are less expensive, simple, flexible, and have better performance. Various saturable absorbers (SAs) have been widely investigated using mode-locking schemes in the passive approach. The construction of the mode-locked operation with an extended cavity fiber laser to generate the desirable nanosecond pulse train has adopted semiconductor saturable absorber mirrors (SESAMs) [4], graphene [5], single-walled carbon nanotubes (SWCNTs) [6], topological insulator (TIs) [7], and transition metal dichalcogenides (TMDs) [8]. However, these SAs have disadvantages and drawbacks that limit their potential as mode-lockers. For SESAMs, the complex fabrication makes it very costly. Apart from that, a limited response time, narrow operating wavelength region, and low damage threshold are also being highlighted as the limitations of SESAMs. As for graphene, its zero-bandgap structure has limited its applications in ultrafast optics. SWCNTs depend on their tube sizes for their absorption efficiency, which determines the bandwidth operation. The intricate fabrication process of TIs is its main flaw, and TMDs' application in the mid-infrared region is limited by large direct bandgaps.

The search for a potential material to provide the next laser generation with low cost, wideband absorption, a high damage threshold, and a reliable SA continues in this work. The selection of suitable SAs and their parameter control are crucial subjects for the mode-locking performance. Amongst many transition metal oxides (TMOs), titanium dioxide (TiO_2) still holds strong attention in various applications, including photonic, photovoltaic, photocatalytic, electrochromic devices, and sensing [9]. Available in three different polymorphs: anatase, rutile, and brookite, TiO_2 possesses many advantageous qualities to offer, such as high chemical and optical stability, nontoxicity, and low cost [10]. Anatase, which is more favored than the other polymorphs, has shown a great deal of interest in solar cells due to the possibility of a higher conduction band edge energy and a lower electron-hole pair recombination rate. In an ultrafast fiber laser, a fast recovery time is an important factor in determining a saturable absorber's pulsing capability. As reported by Elim *et al.*, TiO_2 has a very fast recovery time of 1.5 ps, resulting from the Pauli-blocking principle [11]. Rusdi *et al.* demonstrated the use of a TiO_2 film with a thulium-holmium-doped fiber laser, mentioning that its spectral absorption can be extended to the near-infrared (NIR) region, in which absorption is affected by crystal form and the particle size of TiO_2 [12]. To reaffirm this fact, recent research on TiO_2 has been conducted, with the results remaining significantly promising [13,14]. In addition, with a bandgap of 3.2 eV, anatase TiO_2 has been reported to exhibit significant nonlinear optical response [15], thus making it an ideal candidate as a saturable absorber.

This paper reports a mode-locked EDF laser that uses TiO_2 as a SA. The SA device is incorporated into the cavity of the ring laser to act as a mode-locker that generates nanosecond pulses in the C-band region. The prepared SA's modulation depth, nonsaturable absorption, and saturation intensity were 5%, 37%, and 2.15 kW/cm^2, respectively. The mode-locked operation released pulses with a fundamental frequency of 988 kHz and a pulse width of 230 ns. The center wavelength was recorded at 1560 nm with a 3-dB bandwidth of 1.2 nm. The average output power of 2.17 mW has been obtained and pulse energy of 2.19 nJ was observed under a maximum pump power of 187.04 mW.

2. Titanium dioxide (TiO_2) film preparation and characterization

The TiO_2 powder (anatase, 25 nm, Sigma Aldrich Chemical) and de-ionized water were mixed in a 2:1 ratio to prepare the TiO_2 film fabrication. 30 mg TiO_2 powder and 15 ml of de-ionized water were blended for 3 hours at room temperature using a magnetic hot plate stirrer. The mixture was then placed in an ultrasonic bath for one hour to ensure proper mixing of the TiO_2 powder and de-ionized water. The following step is to combine the TiO_2-deionized water solution and the PVA solution in a 20ml:10ml ratio. The PVA solution was made by dissolving 1 gram of PVA powder in 120 mL of distilled water. Both solutions were stirred for 1 hour with a magnetic stirrer and then placed in a 30-minute ultrasonic bath. Finally, the TiO_2 suspension was put into a petri dish and allowed to dry at room temperature for 48 hours to form a TiO_2 composite film. Figure 1 (a) shows the real image of a TiO_2 PVA film whose thickness was determined to be approximately 50 μm. According to Figure 1 (b), the TiO_2 film's saturable intensity, modulation depth, and non-saturable absorption are 2.15 kW/cm^2, 5%, and 37%, respectively. The film's measurement provided by field emission scanning electron microscopy (FESEM) is depicted in Figure 1 (c). It can be seen from the figure that the TiO_2 appears to be distributed evenly throughout the film. In Figure 1 (d), an absorption of EDFL Q-switched at approximately 13 dB can be seen, which depicts the TiO_2 film's linear absorption profile (LAP) at 1560 nm.

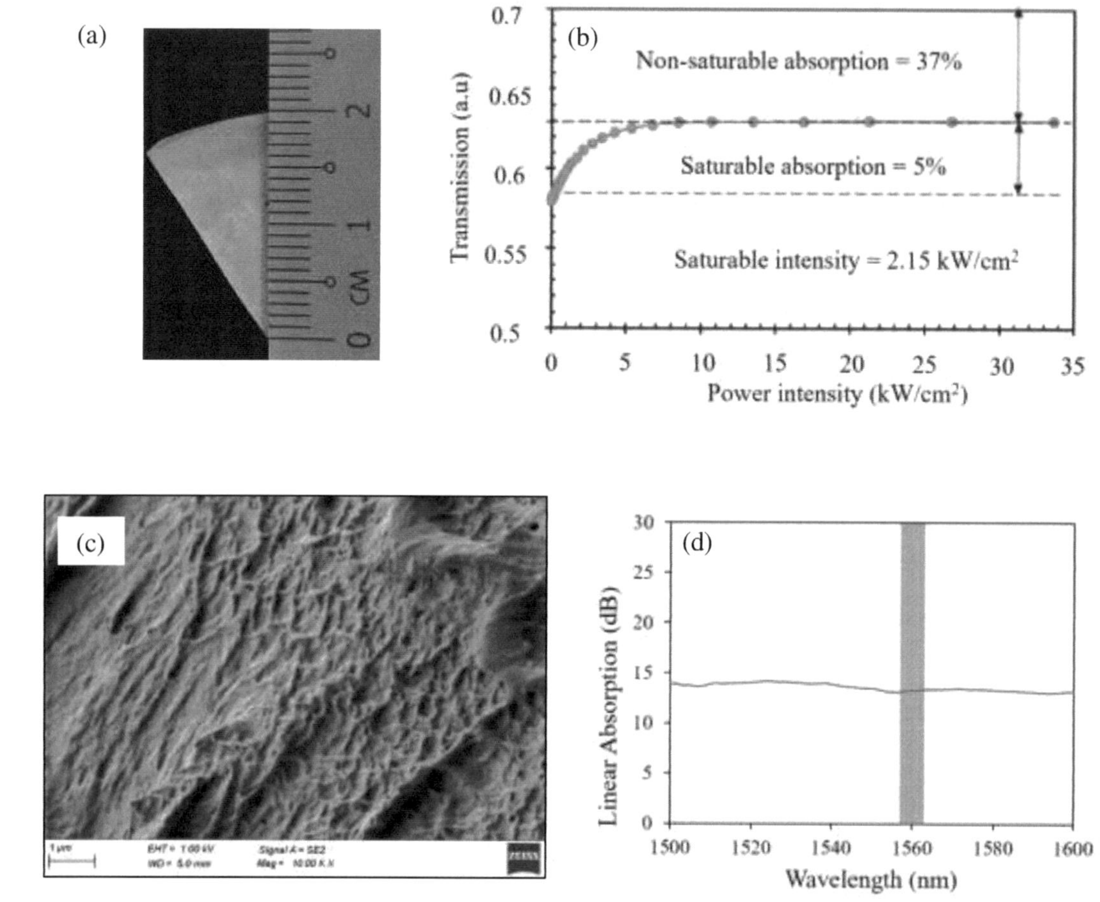

Figure 1. Characteristics of TiO_2 film (a) the real image of the fabricated TiO_2 PVA film (b) nonlinear absorption profile (c) FESEM image (d) linear absorption profile

3. EDFL ring cavity setup

Figure 2 depicts the schematic diagram of the proposed laser cavity for generating the mode-locked EDFL. The laser cavity is composed of a 980 nm laser diode, a wavelength division multiplexer (WDM), a 2.4-meter-long erbium-doped fiber (EDF) gain medium, an isolator, a standard 200-m-long single-mode fiber (SMF), a 90/10 coupler, and the prepared TiO_2 PVA film on the flat end of the fiber ferrule. The 980 nm laser diode forward pumps the erbium-doped fiber (EDF) gain medium via the wavelength division multiplexer's 980/1550 nm port. To prevent unwanted feedback of light in the ring cavity, an optical isolator is used. The light is then directed into the fiber connecter, which is coated with a small piece of TiO_2 PVA film. A fiber-compatible SA device was created by sandwiching the film between two fiber ferrules using a fiber adapter. To ensure proper film attachment and to minimize parasitic reflection, an index matching gel was spread onto the ferrule tip. After passing through the SA, 10% of the laser output is tapped out by the 3dB coupler, while the remaining 90% is retained to circulate within the ring cavity. A 500 MHz digital oscilloscope was used to measure the repetition rate and pulse width while the optical spectrum was observed using a 50 nm spectral resolution optical spectrum analyzer (OSA). After that, the output coupler is connected to a power meter, which records the average output power. The stability of the pulsed laser is verified using a 7.8 GHz RF spectrum analyzer's electrical radio frequency (RF) spectrum signal. The ring cavity's total length was approximately 212 meters.

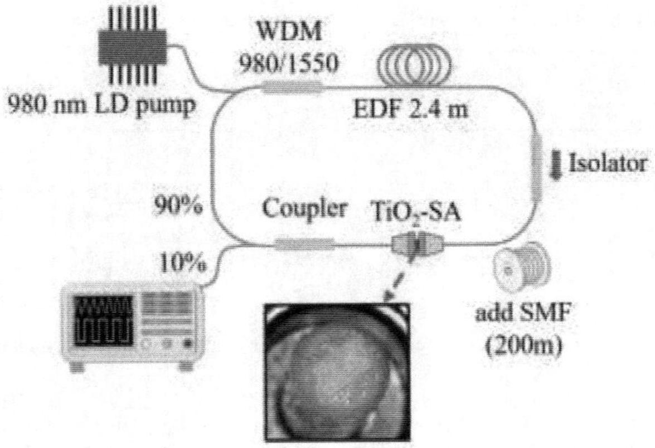

Figure 2. Schematic diagram of Q-switched erbium-doped fiber laser

4. Result and Discussion

The experiment began by varying the pump power at 980nm that was directed into the gain medium via the WDM. The TiO_2-SA and 200m of SMF were incorporated into the ring cavity to compensate for cavity dispersion. The pump power was adjusted over a threshold of 145.83 mW to achieve stable and self-starting mode-locked operation. As a result of increasing the pump power to the maximum allowable value of 187.04 mW, the optical spectrum of mode-locked EDFL was obtained as plotted in Figure 3 (a). As can be seen, the central peak of the wavelength is at 1560 nm with a 3-dB spectral bandwidth of approximately 1.2 nm and a peak power intensity of -2.97 dB. The slight spectral bandwidth is caused by the cavity's self-phase modulation (SPM). Figure 3 (b) shows the mode-locked pulse oscilloscope trace at 187.04 mW pump power. The repetition rate was measured at 988 kHz with a pulse width of 230 ns.

The RF spectrum obtained at maximum pump power is shown in Figure 4, where the pulsed laser's stability was examined further. The figure also included an expanded pulse view that corresponds to a repetition rate of 988 kHz. The harmonic trend indicates that the mode-locking pulses are in the nanosecond range. The RF spectrum shows that the SNR is very high at 69 dB, indicating a stable regime. The measured RF spectrum demonstrates unequivocally that the obtained mode-locked pulses are extremely stable over a wide span of 50 MHz with harmonics up to more than 20th harmonics. To test the stability of the mode-locking operation, an observation was conducted for 24 hours with the pump power maintained at 187.04 mW. The optical spectrum taken every 5 hours at a pump power of 187.04 mW is depicted in Figure 5. The spectra demonstrate how the spectrums are generated in a stable mode-locking regime over a 24-hour period without causing any damage to the TiO_2-PVA thin film. Furthermore, to demonstrate that the TiO_2-SA was responsible for the laser's mode-locking pulse generation, the TiO_2-PVA film was removed from the cavity. At any pump power setting, no pulse was visible on the oscilloscope. This confirms that the prepared film is in charge of the mode-locking operation.

Figure 6 depicts the single pulse energy and output power performances at various pump power inputs. The plot shows that as the input pump power increased from 145.83 mW to 187.04 mW, the output power increased linearly from 1.68 mW to 2.17 mW. The plot shows a slope efficiency of 1.16%. The high insertion loss of the SA caused the relatively low efficiency. Cavity optimization, which is dependent on intra-cavity loss, can be used to improve efficiency. As can be seen in the graph, when the pump power increases, so does the pulse energy. At 187.04 mW pump power, the maximum pulse energy of 2.19 nJ was obtained. These results reinforce that TiO_2 material has the potential to be used as a mode-locker in an EDFL cavity for C-band operation. The proposed TiO_2-based laser's pulse performance was also compared to similar approaches based on other SAs, as shown in Table 1. In comparison to graphene and SWCNT materials, TiO_2 has a higher achievable pulse energy. The TiO_2-

based laser also provides a shorter pulse width, higher repetition rate, and excellent SNR performance when compared to most of the materials in the table.

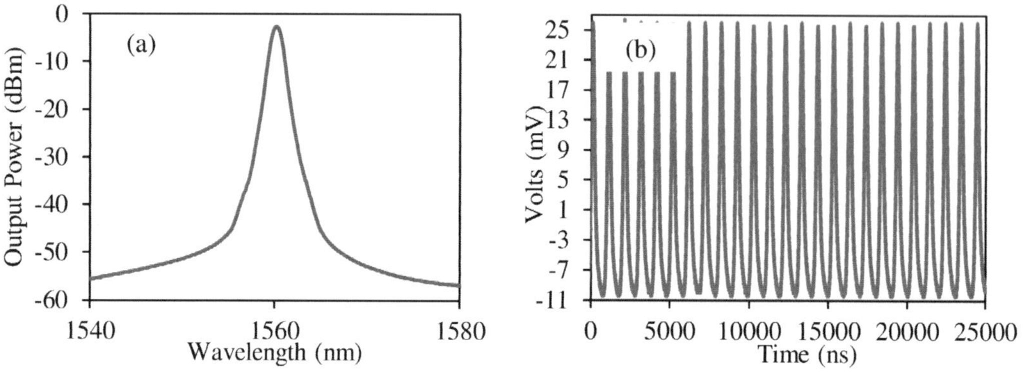

Figure 3. The characteristics of the mode-locking pulses (a) wavelength and (b) pulse train at 187.04 mW

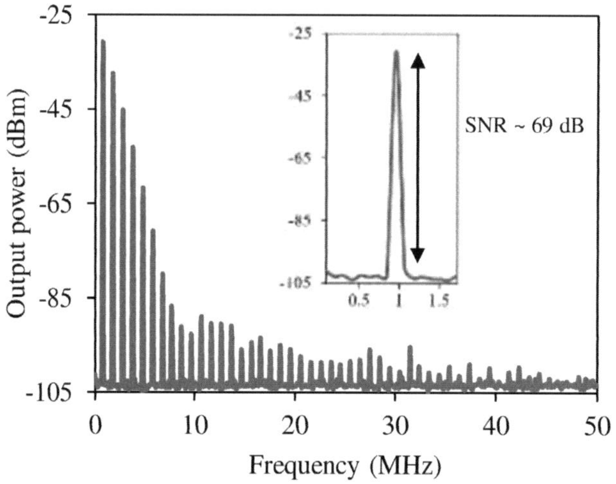

Figure 4. RF spectrum at a pump power of 187.04 mW

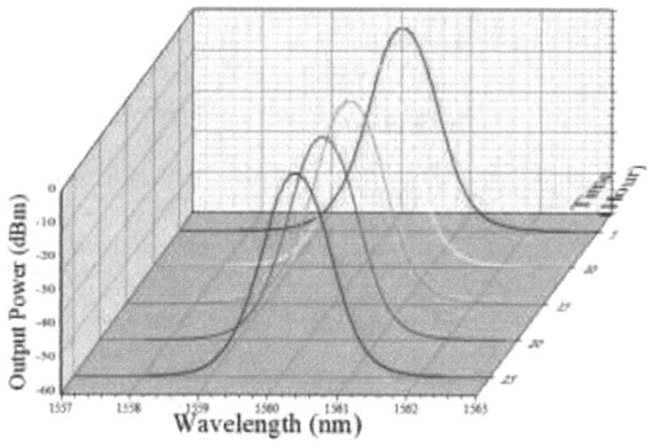

Figure 5. Long-term operation stability of the optical spectrum at the pump power of 187.04 mW

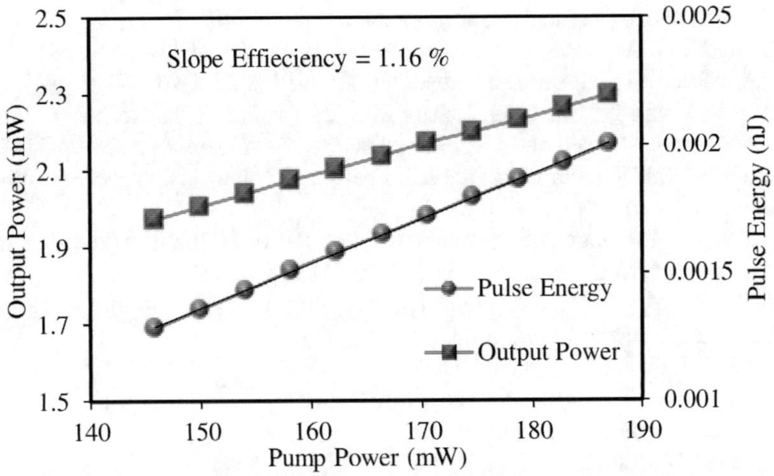

Figure 6. Average output power and single pulse energy vary with the increasing pump power

Table 1. Comparison of the proposed SA's performance to that of other SAs

SA materials	λ (nm)	Threshold (mW)	Pulsewidth (ns)	Repetition rate (kHz)	Max. pulse energy (nJ)	SNR (dB)	Ref.
ZnO	1558.34	45.4	400	1000	11.6	64	[16]
Sm_2O_3	1569.8	109	356	969	4	56	[17]
GaN	1562	145.83	485	967	3.1	59	[18]
SWCNT	1570.4	41.3	332	909	0.34	36	[19]
Graphene	1560	27	6	388	1	45	[20]
TiO_2	1560	145.83	230	988	2.19	69	This work

5. Conclusion

The success of this demonstration led to the conclusion that the TiO_2-SA could generate stable nanosecond pulses in an EDFL that operates at a central wavelength of 1560 nm. The repetition rate recorded was almost constant at 988 kHz within a range of pump power from 145.83 mW to 187.04 mW. The pulse duration was observed at 230 ns. The maximum pump power draws an output power and pulse energy of 2.17 mW and 2.19 nJ, respectively. A very high signal-to-noise ratio (SNR) of 69 dB and a spectral bandwidth of 1.2 nm suggest that TiO_2-SA exhibits comparable performance and gives advantages of stability, simplicity in fabrication, and low cost.

References

[1] Perez-Herrera RA, Diaz S, Fernández-Vallejo M, López-Amo M, Quintela MA, Lopez-Higuera JM. 2009 Switchable multi-wavelength erbium-doped fiber laser for remote sensing *20th Int Conf Opt Fibre Sensors* pp 75031Y

[2] Singla U. Fiber Laser Welding Benefits and Industrial Applications. 2017;5(1):7–10.

[3] Sawisch TJ. 2010 Oral Surgery for the General Practitioner *First Edit. Principles and Practice of Laser Dentistry* pp 93–113

[4] Keller U, Weingarten KJ, Kärtner FX, Kopf D, Braun B, Jung ID, et al. 1996 Semiconductor saturable absorber mirrors (SESAM's) for femtosecond to nanosecond pulse generation in solid-state lasers *IEEE J Sel Top Quantum Electron* vol 2 pp 435–451

[5] Geim AK, Novoselov KS. 2009 The rise of graphene. *Nanosci Technol A Collect Rev from Nat Journals* pp 11–19

[6] Wang YG, Cheng SY, Tai PT, Tang J. 2012 Saturable absorber at 940 nm using single wall carbon

nanotubes deposited by vertical evaporation technique. *Optik (Stuttg)* vol 123 pp 348–351
[7] Zhao C, Zhang H, Qi X, Chen Y, Wang Z, Wen S, et al. 2012 Ultra-short pulse generation by a topological insulator based saturable absorber. *Appl Phys Lett.* vol 101 pp 1–5
[8] Wu K, Zhang X, Wang J, Li X, Chen J. 2015 WS_2 as a saturable absorber for ultrafast photonic applications of mode-locked and Q-switched lasers. *Opt Express* vol 23 pp 11453
[9] Chen X, Mao SS. 2007 Titanium dioxide nanomaterials: Synthesis, properties, modifications and applications. *Chem Rev.* vol 107 pp 2891–2959
[10] Bai Y, Mora-Seró I, De Angelis F, Bisquert J, Wang P. 2014 Titanium dioxide nanomaterials for photovoltaic applications. *Chem Rev.* vol 114 pp 10095–10130
[11] Elim HI, Ji W, Yuwono A, Xue J, Wang J. 2003 Ultrafast optical nonlinearity in PMMA-TiO2 nanocomposites. *Appl Phys Lett.* vol 8
[12] Mohd Rusdi MF, Latiff AA, Paul MC, Das S, Dhar A, Ahmad H, et al. Titanium Dioxide (TiO2) film as a new saturable absorber for generating mode-locked Thulium-Holmium doped all-fiber laser *Opt Laser Technol* vol 89 pp 16–20
[13] Kamarulzaman NH, Salleh H, Dagang AN, Mohd Ghazali MS, Ishak N, Ahmad Z. 2020 Optimization of Titanium Dioxide Layer Fabrication Using Doctor Blade Method in Improving Efficiency of Hybrid Solar Cells *Journal of Physics: Conference Series* vol 854
[14] Ramírez AM, Cattin L, Bernède JC, Díaz FR, Gacitúa MA, Del Valle MA. 2021 Nanostructured tio2 and pedot electrodes with photovoltaic application *Nanomaterials* vol 11 pp 1–8
[15] Iliopoulos K, Kalogerakis G, Vernardou D, Katsarakis N, Koudoumas E, Couris S. 2009 Nonlinear optical response of titanium oxide nanostructured thin films *Thin Solid Films* vol 518 pp 1174–1176
[16] Alani IAM, Lokman MQ, Ahmed MHM, Latiff AA, Harun SW. 2018 A few-picosecond and high-peak-power passively mode-locked erbium-doped fibre laser based on zinc oxide polyvinyl alcohol film saturable absorber *Laser Physics* vol 28
[17] Zulkipli NF, Batumalay M, Samsamnun FSM, Mahyuddin MBH, Hanafi E, Izam TFTMN, et al. 2019 Nanosecond Pulses Generation with Samarium Oxide Film Saturable Absorber. *Chinese Phys Lett* vol 36 pp 074203
[18] Samsamnun FSM, Zulkipli NF, Khudus MIMA, Bakar ASA, Majid WHA, Harun SW. 2019 Nanosecond Pulse Generation with a Gallium Nitride Saturable Absorber. *OSA Contin.* vol 2 pp 134–141
[19] Ismail MA, Harun SW, Zulkepely NR, Nor RM, Ahmad F, Ahmad H. 2012 Nanosecond soliton pulse generation by mode-locked erbium-doped fiber laser using single-walled carbon-nanotube-based saturable absorber. *Appl Opt.* vol 51 pp 8621–8624
[20] Xu J, Wu S, Liu J, Wang Q, Yang QH, Wang P. 2012 Nanosecond-pulsed erbium-doped fiber lasers with graphene saturable absorber. *Opt Communications* vol 285 pp 4466-4469

Investigation on Identification of RRIM Clone Series using Various NIR LED sensor

N.E. Abdullah[1*], M. S. Baddri[1], N.K.Madzhi[1], S.L.M.Hassan[1], I.S.A. Halim[1] and A.A.A. Rahim[1]

[1] School of Electrical Engineering, College of Engineering, Universiti Teknologi MARA, 40450 Shah Alam, Selangor, Malaysia

*Corresponding author's email: noor_ezan@uitm.edu.my

Abstract. The evolution of various rubber tree clone has been introduced by Malaysian Rubber Board since 1928 and it is known as Rubber Research Institute Malaysia (RRIM) clone series. To support this, it is suggested research to investigate the behavior of the clone series with the development from a sensing device with several opted NIR wavelength ranges through the latex samples. An optical sensor prototype has been built consist of four types of NIR LED at selected wavelength and the photodiode works as receptor of the light emission. As transmitter, the NIR sensors are obtain from the local distributor in Malaysia which is RC component. Arduino Uno is use as the processor for processing the data and obtain the output trough LCD. There are five selected clone use which are RRIM 3001, RRIM 2008, RRIM 2007, RRIM 2014 and RRIM 2002. The output result which represented by the voltage and then will be analysed using statistical analysis approach for data analysis. The outcome reveal that the selected rubber tree clone series is able to determine based on the NIR properties.

1. Introduction

Malaysian Rubber Board (MRB) have launched a systematic breeding program for rubber tree clone series named Rubber Research Institute of Malaysia (RRIM) series to support and sustain the yield and timber production [1]. This improvement in rubber sector may provide wide opportunity in Research & Development (R&D) from the government [2]. Due to this plan, MRB has released a few series of rubber trees clone breed that have capability to produce more and high quality of raw latex as well as good timber.

The usage of Near Infra-Red (NIR) nowadays is widely known because of its worldwide application and usage, The variety of NIR wavelength provide multiple options that can be used and suitable for various application. For example, the application of NIR in the food industries for analyzing the nutrient contents in manure samples by Asai et al. NIR optical technique also have been applied in baking industry to analyse raw ingredients for protein content, moisture, particle size, ash, starch damage and colour [3]. Other than that, NIR also can be applied to the design and development of the near infrared (NIR) cameras for industrial purpose [4].

There are several studies that have been carried out in agriculture field in Malaysia especially in rubber industry. For example, NIR is used in identification of white root disease that are caused by indigenous fungal parasites, then the data processed using Artificial Neural Network (ANN) techniques in order to classify a decision of early detection of rubber tree white-root disease [5]. This study uses

ANN as analysis method and achieved 84.4% accuracy. By using optical properties of NIR, the characteristic of the white root disease can be identified at early stage and can be prevented from spread and infect other tree. Color spectrum (RGB) method to determine purity of the latex contain and finally able to determine the rubber clone series via Statistical Packages for Social Sciences (SPSS) analysis [6]. Work done by N.A. Khairuzzaman [7] able to achieve the total accuracy is 84.8% rather than previous study. It is used the combination of visible (VIS), NIR and ultraviolet (UV) to identify the early symptom of white root disease based on changes on leaf color using spectrometer Model Carl Zeiss MCS 600 which is it is the latest Micro Spectroscopy and the data then numerically analyzed using statistical approach to observe the significant findings [7].

However as refer to the Malaysian rubber industry, the problem for identification of the rubber clone breed had become crucial as the difference between clones cannot be identified physically and only scientific test or depends on expertise. The process of determination of rubber breed which may consume more time for decision making if using conventional method. Thus, with the combination research about the optical characteristic of NIR, Arduino Uno and the statistical analysis method will help various application in the agricultural industry especially in rubber clone identification. This also will help to understand the NIR specification base on the characteristic towards the rubber tree clone series. Therefore, it is an inspire for this study to take a step ahead to give more understanding and study to investigate the various type NIR wavelength for identification of rubber tree clone series. It is believed that this study may be a helpful resource as well as provide significant contribution towards society

2. Methodology

There were several steps taken in this research which involving the hardware development consists of the optical sensor and the receiver combination, signaling process to amplify the output, display systems using Arduino to display the output voltage from difference NIR. All the processes have been summarized in the flowchart shown in Figure 1.

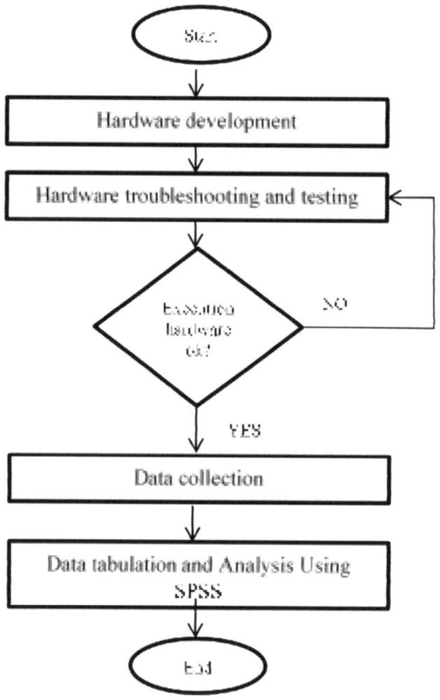

Figure 1. Flowchart of the experiment.

2.1 Sensor and Signal Conditioning Circuit Development

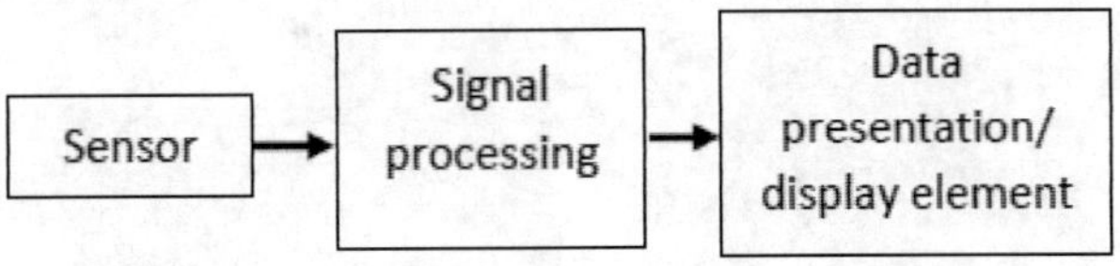

Figure 2. Block diagram of instrumentation system.

Figure 2 shows the overall instrument system that had been used to the combination of three major part which are sensor, signal processing/signal processor and data presentation /display element. Each part play an important role to deliver the output in term of voltage.

Meanwhile the following Figure 3(a) and 3(b) show the sample of NIR LED used and Table 1 show the characteristics of the NIR LEDS as LED sensors with four difference wavelength which is ranging from 850nm up to 950nm. This value is chosen due to the availability of this NIR LEDS in the market. The sensor circuit is design using Proteus 8 software with contain four difference type of NIR LEDS, four LEDS indicator to indicate which NIR is use during the data taken and four switch to control on/off of the NIR. The named for each NIR LEDS is denoted according to Table 1.

3(a) 3(b)

Figure 3. Example of NIR and photodiode use in this study.

Table 1. NIR LED characteristics

Name	Characteristic			
	Wavelength (nm)	*Half Angle (ɵ)*	*Radient Intensity (mW/sr)*	*Power Dissipation (MV)*
NIR1	850	22	50	150
NIR2	870	10	120	180
NIR3	940	25	120	220
NIR4	950	23	120	170

The optical sensor circuit is the improvise version from the previous study by researcher team on identification rubber clone using optical properties [8]. This circuit use LM 324 as the processing element. LM 324 contain buffer and non-inverting amplifier that will amplified the output which detected by the photodiode SFH 229. This photodiode capable to detect NIR wavelength from 380nm to 1100 nm. Figure 4 show the Arduino Uno and the LCD display 16X2 is use for data representation. LCD will display which NIR will be use and the output voltage. Arduino Uno have the ADC (analog to digital converter) to convert the analog signal to digital signal for display purpose [9].

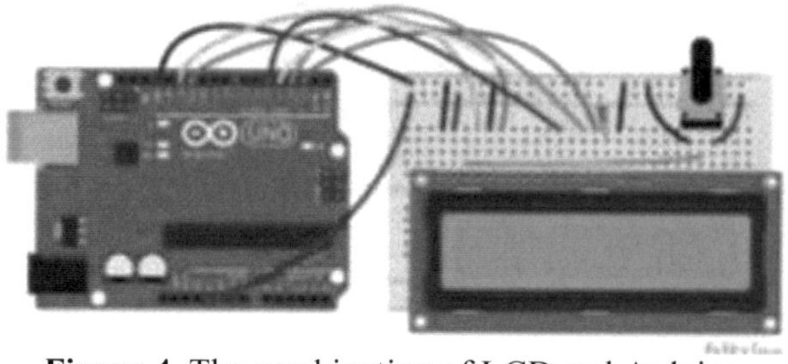

Figure 4. The combination of LCD and Arduino

2.2 Schematic Configuration Setup

Figure 5 represent the construction that have been improvised from the previous studies and suitable for this work. Method of where the sample is placed in the black box to prevent the interference of other optical wavelength that might present during the data observation and collection. The distance of each NIR from the photodiode is 1.5cm. The height of the sample can adjust between 3 cm and 4 cm, according to some circumstances during the experimental work.

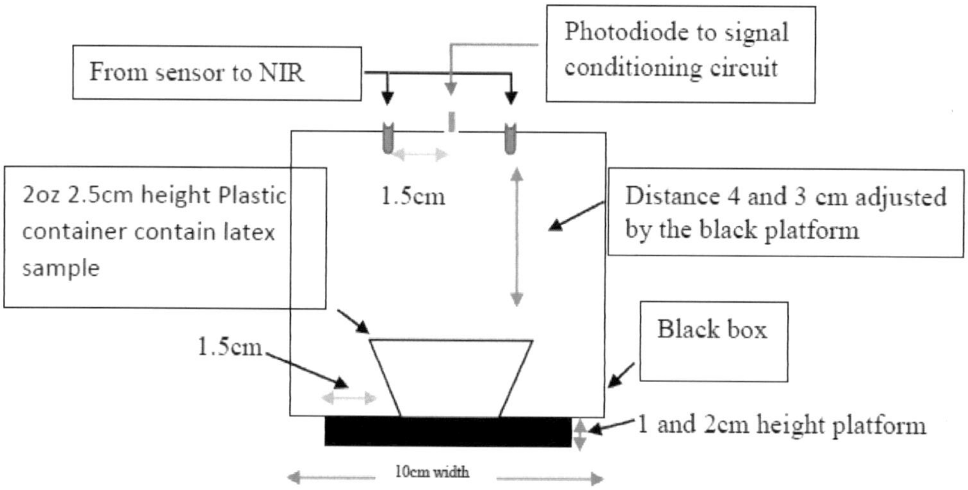

Figure 5. Location of the sample and NIR inside black box 10cm height x 10cm width x 20cm length

Figure 6 display the whole instrument chart is combine and ready to be tested for functionality and troubleshooting purposes.

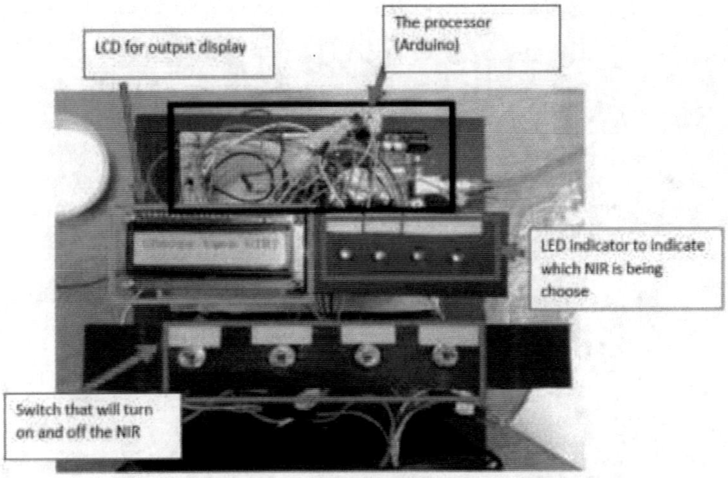

Figure 6. Hardware development on progress

Figure 7 show the principal reflection, scattering and refection of NIR light emission towards on a surface of latex sample. Photodiode then will collect the reflection of the NIR light. Latex sample from difference rubber clone have difference concentration that will cause difference output voltage displayed. Distance sample from NIR will also affect the photodiode absorption due to the each of NIR have difference half angle for NIR emission detection [10].

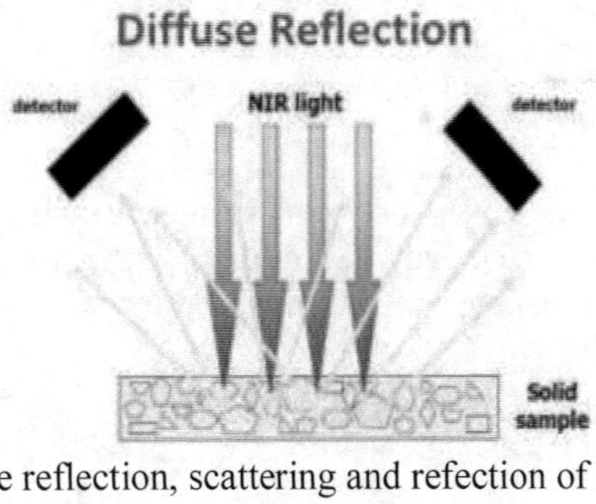

Figure 7. The reflection, scattering and refection of NIR light

While Figure 8 shows the LED emission of a normal modal. The other major reason to choose NIR other than LED emission in spectrum is because the NIR wavelength is suitable for latex composition [11]. The following Figure 9 shows the white and black card which is used during the circuit testing and troubleshooting. Another element used in data calibration before the data collection is distilled water. The purpose of this process is to calibrate the circuit and make some changes if there is any problem. Material such as card box doesn't reflect the NIR emission as the reading output usually still in millivolt, while the plain water reading have quite significant value due to the emission can reflect more on water. This prove the functionality of the circuit before the clone sample is use. Table 2 shows the result for calibration purpose.

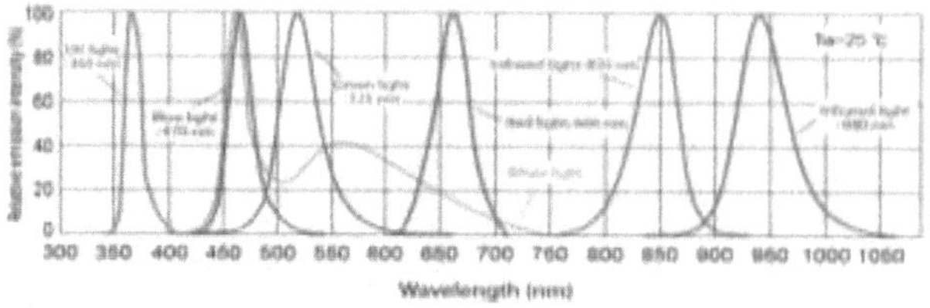

Figure 8. LED emission spectrum distribution for a standard model for difference wavelength

Figure 9. Black and what board use to test the reflection of NIR during circuit development.

Table 2. Calibration result of the used NIR.

Samples	**The output voltage V**			
	NIR1	***NIR2***	***NIR3***	***NIR4***
White	0.31	0.27	0.21	0.23
Black	0.27	0.19	0.21	0.23
Distilled water	2.54	2.23	2.29	2.31

2.3 Sample and Data Collection

The latex sample of RRIM 3001, RRIM 2008, RRIM 2007, RRIM 2014 and RRIM 2002 is used in this experiment to study and observe the NIR performance. The samples were collected from the Rubber Research Institute Malaysia at Kota Tinggi, Johor during high yielding period which falls on November to February [12]. The total 1500 sample have been collected for all clone.

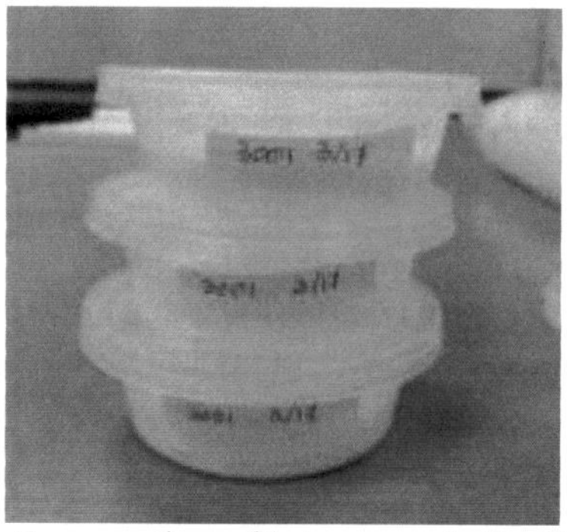

Figure 10. Preservation of latex sample for RRIM 3001

Figure 10 shows on how the latex sample stored in specimen cups. The specimen cup then stored in freezer to preserve the quality of the latex, under controlled environment and can be used for other analysis in future [13].

One of the reasons choosing Infrared light's in this study is because it have long wavelength which often allows for less scattering from part surfaces and higher transmission rates through them. Besides a higher transmission means it can pass through more materials which is show that infra –red (IR) light ideal for identification of the breed

Table 3. Output test based on Distance.

clone	*NIR1*		*NIR2*		*NIR3*		*NIR4*	
	3CM	*4CM*	*3CM*	*4CM*	*3CM*	*4CM*	*3CM*	*4CM*
2002	3.15	3.07	3.03	2.99	3.13	3.04	3.13	3.04
2007	3.15	3.08	3.02	2.99	3.14	3.03	3.12	3.03
2008	3.20	3.13	3.07	3.03	3.17	3.08	3.17	3.10
2014	3.20	3.12	3.08	3.03	3.16	3.08	3.17	3.09
3001	3.23	3.14	3.10	3.04	3.19	3.08	3.18	3.10

Table 3 shows the output voltage result tested using difference distance to obtain the most accurate highest reading. The highest reading from all clones is using 3cm distance. Therefore, 3cm distance is use for further data collection for analysis cause the output voltage is higher than 4cm. Hence, further analysis can be done by using 3cm measurement distance

The output result that obtained from the experiment will be analyzed using Statistical Package for the Social Science (SPSS) software for empirical hypothesis so that the study can be proven with the strong analysis to support. SPSS is chosen because this software stored the data in the spreadsheet condition that look alike Microsoft Excel and can generate routine descriptive statistical data. Statistical analysis that will be carried out for this study is normality test, Error plot and Analysis of Variance (ANOVA) Test approach.

3. Results and Discussion

Table 4 are the behavior of the data distribution according to the maximum and minimum reading obtained for each of the NIR output.

Table 4. Data behavior for Output voltage.

	N	**Minimum**	**Maximum**
NIR1	1500	3.11 V	3.23 V
NIR2	1500	3.01 V	3.11 V
NIR3	1500	3.06 V	3.20 V
NIR4	1500	3.08 V	3.21

3.1 SPSS Analysis

Statistical analysis is the important preliminary analysis should be conducted before proceeding with other analysis. This research recommended several relevant analysis such as the Normality Test for testing the population behaviour whether it is normal or not normal distribution. If the population is normal then the Parametric Test (error plot, T Test, ANOVA test etc.) is suggested if not the Non-Parametric Test (Man –Whittney, Kruskal Wallis Test etc.) is the alternative [14,15].

Journal of Physics: Conference Series **2075** (2021) 012014 doi:10.1088/1742-6596/2075/1/012014

3.1.1 Normality Test Analysis.

SPSS analysis approach is used to investigate the performance of the NIR during data collection. The purpose of the analysis is to understand the behavior of the data whether it's normally distributed or not. This study using Kolmogorov-Smirnova (K-S test) to determine the normality. If the sig (p-value) value more than 0.05, then the population is normal. If not, it will be not normal and the data would be non-significant thus it can't be use or the suitable analysis can be done via Non-Parametric Test [14,16].

Table 5. Test of Normality of Rubber Tree Clone Series

CLONE		Kolmogorov-Smirnov[a]		
		Statistic	*df*	*Sig.*
NIR1	RRIM2002	.032	300	.200*
	RRIM2007	.031	300	.200*
	RRIM2008	.047	300	.200*
	RRIM2014	.048	300	.095
	RRIM3001	.057	300	.021
NIR2	RRIM2002	.030	300	.200*
	RRIM2007	.040	300	.200*
	RRIM2008	.027	300	.200*
	RRIM2014	.037	300	.200*
	RRIM3001	.044	300	.200*
NIR3	RRIM2002	.043	300	.200*
	RRIM2007	.042	300	.200*
	RRIM2008	.035	300	.200*
	RRIM2014	.029	300	.200*
	RRIM3001	.030	300	.200*
NIR4	RRIM2002	.037	300	.200*
	RRIM2007	.036	300	.200*
	RRIM2008	.032	300	.200*
	RRIM2014	.022	300	.200*
	RRIM3001	.052	300	.045

*df:degree of freedom
*sig.:significant value (p value)

Table 5 shows the normality test on each NIR towards the rubber tree clone series. NIR2 and NIR3 show a normal distributed data which can be accepted for further analysis. While NIR1 and NIR4 show un-normal data distribution, thus this implies that NIR1 and NIR4 cannot be used for identification purpose due significant value on RRIM3001 less than 0.05. Hence the parametric analysis only applicable to NIR2 and NIR3.

3.1.2 *Error Plot.*

The first analysis on identification of rubber tree clone series is on using error bar plot. The error plot shows the initial overview graphical analysis on the results. NIR2 and NIR3 then analysed using error plot toward the regularity five rubber clone of latex for further identification

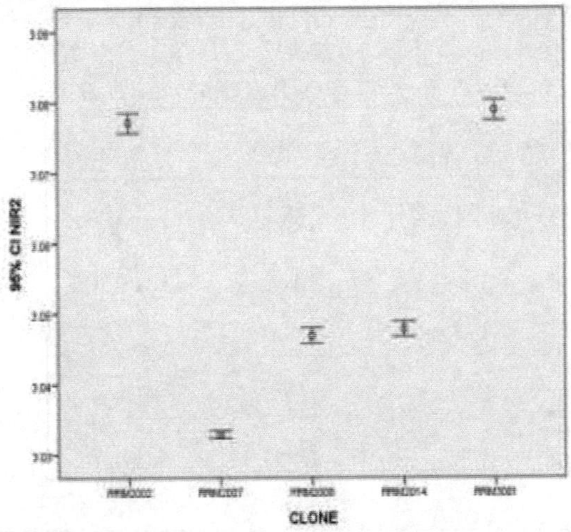

Figure 11. Error plot NIR2 towards five clone.

From Figure 11 represent NIR2 for which only can be differentiated clearly RRRIM2002, RRIM2007 and RRIM3001. However, between RRIM2008 and RRIM2014 the output result is about the same, which conclude that NIR2 cannot be used for identification because it is could not differentiate well for clone RRIM2008 and RRIM2014.

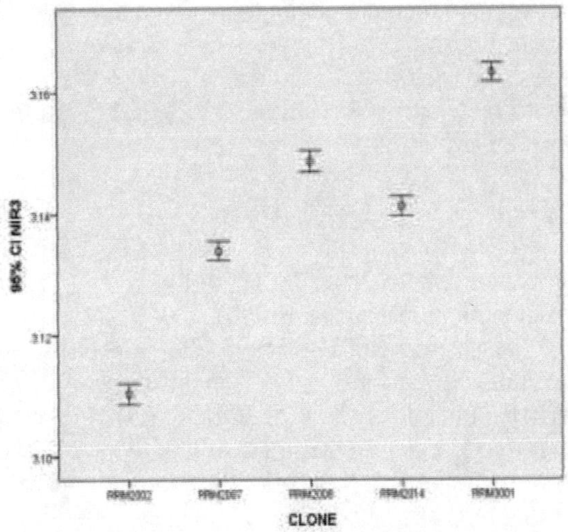

Figure 12. Error plot NIR3 towards five clone

Hence Figure 12 represent analysis for NIR3 which perform very well in which each of clone can be measured significant difference between each other. Thus NIR3 is suggested to use for rubber clone identification purpose. Hence ANOVA test being conducted to enhance finding done by Error Bar Plot

3.1.3 ANOVA Test Analysis.

The following results are based on the numerical value. This is to proven the previous graphical observation using Error Bar Plot. The ANOVA Test suitable for more than three means to be analysed [14]. As can be seen that Table 6 tabulated the ANOVA output analysis. By this method, NIR2 and NIR3 can be used to identify five clone series due to the fact that the sig value is less than 0.05. Thus this implies that rubber tree clone can be identified using both NIR.

Table 6. ANOVA Test Output Result.

		Sum of Squares	df	Mean Square	F	Sig.
NIR2	**Between Groups**	0.496	4	0.124	1137.278	0.000
	Within Groups	0.163	1495	0.000	-	-
	Total	0.659	1499	-	-	-
NIR3	**Between Groups**	0.462	4	0.115	536.151	0.000
	Within Groups	0.322	1495	0.000	-	-
	Total	0.783	1499	-	-	-

4. Conclusion

The process of identification the numerous rubber clone series is an impossible if it does not depending on the expertise. Hence by this research, the types of clones can be identified using an electrical approach and it could help any individuals and institution to recognize the clone types according to the numerical information. It can be concluded that the difference type of NIR might cause of difference output. This is supported by the datasheet provided by the manufacturer. Even though the NIR have high wavelength but other aspect such as forward current, half angle, relative radiant intensity might affect the NIR performance. Thus, it can be concluded that the investigation process is successfully able to identify the suitable NIR LED sensor to recognize RRIM clone series and it can be used to support the future works in developing sophisticated system.

References

[1] Jenet, The Malaysian Natural Rubber Industry, (2008)
[2] Jenet; Rubber Tree Sustainable in Malaysia. (2008)
[3] Asai, T., S. Shimizu, T. Koga, and M. Sato. 1993. Quick determination of total nitrogen, total carbon, and crude ash in cattle manure using near-infrared reflectance.
[4] FLIR.R&D and industrial applications for Near Infrared (NIR) camera. Technical note (2011)
[5] Sulaiman, M. S., et al. (2014). "Early Detection of White Root Disease for Rubber Tree based on Leaf Discolouration with Neural Network Technique" Journal of Applied Science and Agriculture (JASA) 9(11): 341-351.
[6] Hadzli Hashim, Noor Ezan Abdullah, Suhaimi Sulaiman, Marwani Ismail, "Classification of Hevea Brasiliensis Leaf Diseases through Imaging Technique and Neural Network". J. Appl. Sci. &Agric., 9(11): 358-365, 2014.
[7] Noor Aishah Khairuzzaman, Hadzli Hashim, Nina Korlina Madzhi, Noor Ezan Abdullah, Faridatul Aima Ismail, Color Spectrum Properties of Pure and Non-pure Latex in Discriminating Rubber Clone Series. Faculty of Electrical Engineering, Universiti Teknologi MARA,

[8] Noor Ezan Abdullah1, Hadzli Hashim, Nina Korlina Madzhi, Mohd Suhaimi Sulaiman, Faridatul Aima Ismail,Kamaruddin Mohd Hanafia ,Rubber Tree Clone Breed Identification based on Latex Spectrum Properties .2015. Trans Tech Publications, Switzerland

[9] Simon Monk. Arduino Lesson 11. LCD Displays. © Adafruit Industries Last Updated: 2013-06-11

[10] Dahm, Donald J., and Kevin D. Dahm. "The Physics of Near-Infrared Scattering." Near-Infrared Technology in the Agricultural and Food Industries. Ed. Karl Norris, and Phil Williams. St. Paul, MN: American Association of Cereal Chemists, Inc., 2001. 1-18.

[11] Shenk, John S., et. al. "Application of NIR Spectroscopy to Agricultural Products." Handbook ofNear-Infrared Analysis. Ed. Donald A. Burns, and Emil W. Ciurczak. New York: Marcel Dekker, Inc., 2001. 419-474.

[12] Faridatul Aima Ismail1, Hadzli Hashim, Nina Korlina Madzhi, Noor Ezan Abdullah, Azri Faris Mohd Azmi, Noor Aishah Khairuzzaman, Mohd Afiq Asyraf Mohd Nazi, Nur Addina Balqis Adnan. Identification of Rubber Clone Series through Color Spectrum for Various Moisture Content. Faculty of Electrical Engineering Universiti Teknologi MARA.

[13] Jayanthy, T, & Sankaranarayanan, PE. (2005). Measurement of dry rubber content in latex using microwave technique. Measurement Science Review, 5(3), 50-54

[14] Haron, H. (2010). Understanding Statistics with SPSS: A friendly Approach, University Publication Centre (UPENA), UiTM

[15] Brown, R. B. and M. Saunders (2007). Dealing with statistics: What you need to know, McGraw Hill Education.

[16] Ralph B. D'Agostino (1986). "Tests for the Normal Distribution". In D'Agostino, R.B. and Stephens, M.A.. Goodness-of-Fit Techniques. New York: Marcel Dekker. ISBN 0-8247-7487-6.

Acknowledgments
The authors would like to express their gratitude for several individuals and agencies that have been involved during the research works. Special thanks to College of Engineering, Universiti Teknologi MARA for supporting this work. Also, not to forget to research grant 600-RMC/GPK 5/3 (032/2020).

Random laser behavior in Gold-doped Zinc Oxide nanorods structures

Abdullah Taha Ali[1], W. Maryam[1*], Yu-Wei Huang[2], H.C. Hsu[2], Naser M. Ahmed[1], N. Zainal[3] and H. Abu Hassan[1]

[1] School of Physics, Universiti Sains Malaysia, 11800 Penang, Malaysia
[2] Department of Photonics, National Cheng Kung University, Tainan, 70101, Taiwan.
[3] Institute of Nano Optoelectronics Research and Technology (INOR), Universiti Sains Malaysia, 11800, Penang, Malaysia.

[*]Corresponding author's email: wanmaryam@usm.my

Abstract. The random laser was investigated in gold-doped Zinc Oxide nanorods (Au-doped ZnO NRs) under a range of pumping power 0.25 - 4.66 mW. The Au-doped ZnO NRs prepared by chemical bath deposition (CBD) on the ZnO seed layer, were pre-coated on glass substrate using radio frequency magnetron sputtering (Rf-sputtering). The morphological of Au-doped ZnO NRs shows a hexagonal and strong vertically alignment against the substrate. The Energy dispersive spectroscopy (EDX) spectrum and elemental mapping results confirmed that Au atoms (at.%) are doped and spread over the ZnO NRs. More interestingly, the random laser of Au-doped ZnO shows a redshift of ~38 nm. This study showed the ability of using doping as a tuning parameter in the random laser, also provided an emphasis on Au-doped ZnO NRs as suitable options for controllable random laser devices.

Keywords: Doped ZnO nanorod; tunable emission; photonic devices; nanotechnology

1. Introduction

In recent decades, the random lasers has potential in many industrial and medical applications [1,2], due to their simple (mirrorless) structure [2,3]. This contrasts with conventional lasers, where the medium is placed in a custom optical cavity with a mirror. Random lasing phenomena are induced by multiple photons scattering in a disordered medium [4]. Threshold power, lasing tunability and directionality of guided wave plays an important role in determining a random laser characterization. On the other hand, a randomness of multi-scattering hinders the random laser from being easily controlled. A considerable amount of research on tunable random lasers has been conducted to move random lasers from theory to practical application [5–10]. Nonetheless, most reported research showed that the intensity of a random laser decreases as the temperature or applied voltage increased, which is far from practical use. in addition, most tuning parameters are external controls, like temperature, voltage, or electromagnetic wave, etc., not considered as suitable controls in terms of the embedded optoelectronic devices (like- photodetectors). However, despite these difficulties, the tuning random laser using doping technique is more reliable in terms of low-cost and applicability.

Random laser has been investigated in different nanostructure materials, such as, ZnO [11], GaAs [12], ZnSe [13], as well as organic molecules [14,15] have been used as scattering medium for random lasing. On the other hand, the simplicity and low-cost fabrication techniques of the nanostructured materials, such as ZnO profile, bringing a random laser as an attractive candidate for employment in a various range of light-emitting devices [1].

The UV lasing induced by ZnO nanostructures has stimulated a great deal of interest due to its wide range applications for UV devices. As well as large bandgap (3.37 eV) and having high excitonic binding energy (60 meV) [16–18], the ZnO considered is an attractive lasing material, and the hexagonal ZnO nanorods (NRs) with high crystallinity and the ability to doped with group-IB elements, could be fabricated in simple methods, such as chemical bath deposition [19–22].
In this paper, the random laser investigated by Au-doped ZnO nanorods were fabricated by low-cost method (CBD), under a sequence of pumping power to study the threshold and tunability of the random laser emissions.

2. Experimental procedure

Au-doped ZnO NRs were synthesized via CBD. A thin film (100 nm) of ZnO seed layer was coated on glass slide (25mm×10mm×1.2 mm) by RF-magnetron sputtering. Then, the samples were heated in the furnace at 350°C for 1 hour under air ambient. Next, an aqueous solution of 1:1 (0.08M) zinc nitrate [$Zn(NO_3)_2.6H_2O$] and Hexamethylenetetramine [HMT; $C_6H_{12}N_4$] was stirred for half an hour. At the same time, another aqueous solution of 0.008 M of gold chloride [$HAuCl_4.4H_2O$] were stirred separately for 30 minutes. Then, both solutions were mixed, and stirred for a further 15 minutes and poured into semi autoclave glass bottles. The sample was positioned 45° faced down against bottle wall, And the caps were secured tightly. Afterwards, the glass bottles were left in an oven already preheated at 80°C for 4 hours. Finally, the samples were collected from the glass vials, rinsed with distilled water multiple times, and then slowly dried with high-purity nitrogen.

The morphological and compositional variations of the samples were determined using field emission scanning electron microscope (FESEM; Nova NanoSEM 450, FEI, Japan) coupled with energy dispersive X-ray (EDX). The random laser measurements were carried out using a micro-PL system equipped with a Nd:YAG pulsed laser source operating at 355 nm, 1 KHz repetition rate, and pulse width of 350 ps. The sample holder was mounted on a three-axis translational stage to concentrate the laser light on the sample, which was viewed through a camera positioned at an angle from the sample holder. The laser emission was captured by optical cable that was linked to a spectrograph with a liquid-N2 CCD array detector. All tests were carried out at room temperature, and the beam spot size set at about 100 μm in diameter.

3. Results and discussion

FESEM top view and cross section as-grown Au-doped ZnO NRs are presented in figure 1(a-b). As observed, the average nanorods diameter of doped sample is around 380 nm while the average length is about 1.85 μm. The surface morphology of the Au-doped ZnO NRs has a hexagonal shape and is vertically aligned against the surface of the substrate. However, the Au-doping has no discernible effect on the morphology of ZnO nanorods. Nonetheless, compositional analysis of the sample using EDX (figure 1(d)) confirms the presence of elemental gold (< 0.2 %atomic) in ZnO NRs. This low atomic percentage is due to the relatively low concentration of gold chloride in the precursor solution as compared to zinc nitrate during the synthesis. Figure 1(c) obtain corresponding Zn, O and Au EDX mapping images. These EDX mapping images reveal that Au atoms were distributed uniformly inside ZnO NRs.

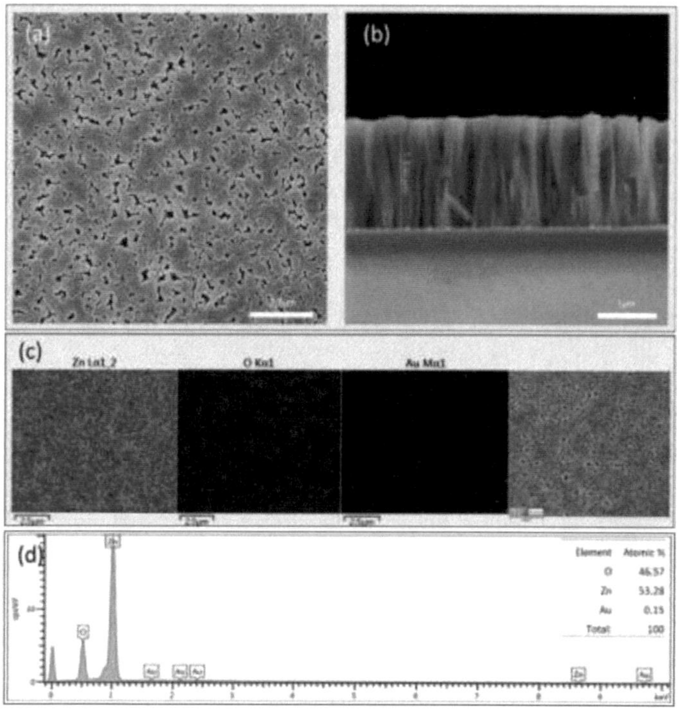

Figure 1: (a-b) top view and cross section of Au-doped ZnO NRs , (c) Elemental mapping images of Au-doped ZnO nanorods sample, and (d) EDX results showing the atomic percentage of elements in the sample.

Figure 2 shows the lasing emission with different pumped powers applied in the Au-doped ZnO NRs samples. The minimum lasing threshold value was 1.13 mW. Also, we founded the intensity of lasing peaks was increased upward against decreasing FWHM with increasing pump power (figure 2b). Interestingly, when the doped sample is pumped above 4 mW, the lasing peak shifted to ~38 nm, towards the visible wavelength range. The incorporation of the foreign atoms into a semiconductor structure acts to broaden the valance band, which in turn leads to narrow or contraction of the bandgap states [23]. As a Group-IB element, the occupied d orbital energies of Au are close to the oxygen p level, and both the O p and Au d orbits exhibit t2 symmetry in the tetrahedral environment. [19]. When the Au atom is doped in Zn, the strong p-d coupling between Au and O occurs, raising the O 2p level up and reducing the direct fundamental bandgap. Therefore, the narrowing bandgap in the Au-doped ZnO structure leads to the shifting of lasing peaks towards the higher wavelengths. But in our study, the shifting occurs only at high pumping powers. Ryo Niyuki *et al*, reported for the first time, a red-shift of lasing peak founded in ZnO nanoparticles when the excitation intensity increased above 10 MW/cm^2 [24]. This lasing peak transition was explained due to the involvement of exciton recombination. However, with high enough excitation power, and with existing Au atoms doped in ZnO structure, which might be causing more exciton recombination, especially at the O 2p level that affected by Au d orbit, and the transition to lasing photon with population inversion will occur at the lowered energy levels, thus the lasing peaks transitions to the higher wavelength regions.

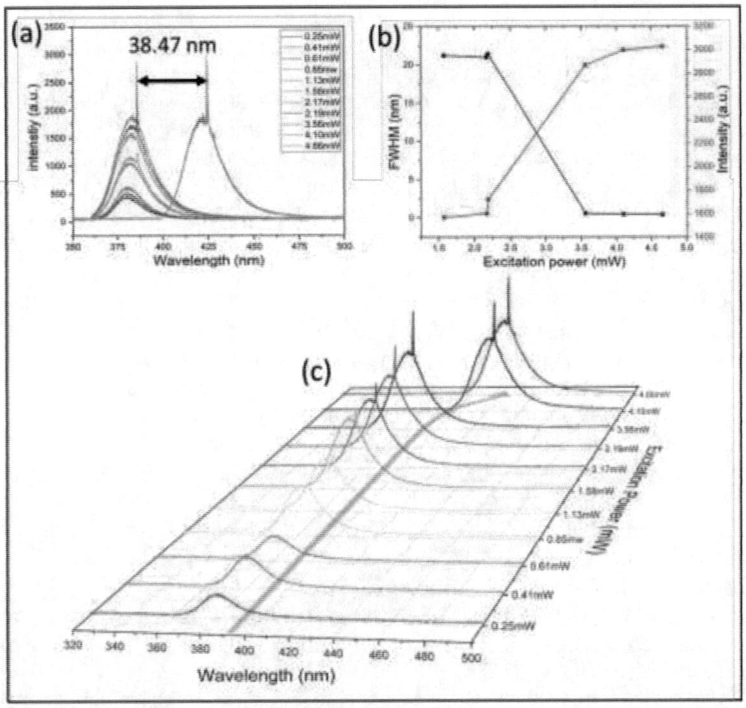

Figure 2: (a and c) Optical pumping power dependence lasing measurement from Au-doped ZnO NRs, and (b) FWHM and the highest intensity lasing mode that appeared at different pump powers

4. Conclusion

Au-doped ZnO NRs were synthesized via (CBD) at 80°C growth temperature. The hexagonal Au-doped ZnO NRs deposited on glass substrates were vertically well aligned. The compositional analysis of the sample using EDX elemental gold (< 0.2 %atomic) in ZnO NRs and the EDX mapping images reveals that Au atoms were distributed uniformly inside ZnO NRs. The minimum random lasing threshold value is 1.13 mW, and we noticed that the intensity of lasing peaks was increased upward versus decreasing of FEWM with increasing the pumping power. Moreover, changing the power pumping, shifted wavelength, giving tunability about ~38 nm. This study showing that Au-doped ZnO NRs is the promising candidate for controlling the random lasing device.

References

[1] D.S. Wiersma, The physics and applications of random lasers, Nat. Phys. 4 (2008) 359–367. https://doi.org/10.1038/nphys971.
[2] F. Luan, B. Gu, A.S.L. Gomes, K. Yong, S. Wen, P.N. Prasad, Lasing in nanocomposite random media, Nano Today. (2015) 1–25. https://doi.org/10.1016/j.nantod.2015.02.006.
[3] H. Cao, Lasing in random media, Waves in Random Media. 13 (2003) R1–R39.
[4] H. Ramachandran, Mirrorless lasers, (2002) 313–323.
[5] C.-R. Lee, S.-H. Lin, J.-W. Guo, J.-D. Lin, H.-L. Lin, Y.-C. Zheng, C.-L. Ma, C.-T. Horng, H.-Y. Sun, S.-Y. Huang, Electrically and thermally controllable nanoparticle random laser in a well-aligned dye-doped liquid crystal cell, Opt. Mater. Express. 5 (2015) 1469. https://doi.org/10.1364/OME.5.001469.
[6] S.Y. Lin, T.M. Sun, C.S. Wang, C.S. Liao, P. Perumal, C.W. Chiang, Y.F. Chen, Stretchable random lasers with tunable coherent loops, Opt. InfoBase Conf. Pap. Part F68-L (2017). https://doi.org/10.1364/LS.2017.LTh3F.3.
[7] Y.J. Lee, T.W. Yeh, Z.P. Yang, Y.C. Yao, C.Y. Chang, M.T. Tsai, J.K. Sheu, A curvature-tunable random laser, Nanoscale. 11 (2019) 3534–3545. https://doi.org/10.1039/c8nr09153f.
[8] S.F. Haddawi, H.R. Humud, S.M. Hamidi, Tunable low power piezo-plasmonic random laser

under external voltage, Optik (Stuttg). 207 (2020) 164482. https://doi.org/10.1016/j.ijleo.2020.164482.

[9] X. Shi, Q. Chang, Y. Bian, H. Cui, Z. Wang, Line Width-Tunable Random Laser Based on Manipulating Plasmonic Scattering, ACS Photonics. 6 (2019) 2245–2251. https://doi.org/10.1021/acsphotonics.9b00508.

[10] X. Meng, J. Ma, K. Xie, L. Hong, J. Zhang, Z. Hu, Tunable random laser in flexible hydrogel, Opt. Mater. (Amst). 115 (2021) 111027. https://doi.org/10.1016/j.optmat.2021.111027.

[11] H. Cao, Y.G. Zhao, S.T. Ho, E.W. Seelig, Q.H. Wang, R.P.H. Chang, Random Laser Action in Semiconductor Powder, Phys. Rev. Lett. 82 (1999) 2278–2281. https://doi.org/10.1103/PhysRevLett.82.2278.

[12] T. Nakamura, T. Takahashi, S. Adachi, Temperature dependence of GaAs random laser characteristics, (2010) 1–6. https://doi.org/10.1103/PhysRevB.81.125324.

[13] T. Takahashi, T. Nakamura, S. Adachi, Blue-light-emitting ZnSe random laser, Opt. Lett. 34 (2009) 3923. https://doi.org/10.1364/OL.34.003923.

[14] F. Quochi, F. Cordella, A. Mura, G. Bongiovanni, F. Balzer, H.-G. Rubahn, One-Dimensional Random Lasing in a Single Organic Nanofiber, J. Phys. Chem. B. 109 (2005) 21690–21693. https://doi.org/10.1021/jp054324r.

[15] Y. Wang, Z. Duan, Z. Qiu, P. Zhang, J. Wu, D. Zhang, T. Xiang, Random lasing in human tissues embedded with organic dyes for cancer diagnosis, Sci. Rep. 7 (2017) 8385. https://doi.org/10.1038/s41598-017-08625-3.

[16] H.-C. Hsu, C.-Y. Wu, H.-M. Cheng, W.-F. Hsieh, Band gap engineering and stimulated emission of ZnMgO nanowires, Appl. Phys. Lett. 89 (2006) 013101. https://doi.org/10.1063/1.2218813.

[17] A.S. Gadallah, K. Nomenyo, C. Couteau, D.J. Rogers, G. Lérondel, Stimulated emission from ZnO thin films with high optical gain and low loss, Appl. Phys. Lett. 102 (2013) 1–5. https://doi.org/10.1063/1.4803081.

[18] W.M. Wan Ahmad Kamil, M.K. Shakfa, M.N. Nordin, D. Priante, Lasing from ZnO nanorods on ITO-coated substrates, (2018) 27. https://doi.org/10.1117/12.2307131.

[19] Y. Yan, M.M. Al-Jassim, S.H. Wei, Doping of ZnO by group-IB elements, Appl. Phys. Lett. 89 (2006). https://doi.org/10.1063/1.2378404.

[20] Q. Li, J. Bian, J. Sun, J. Wang, Y. Luo, K. Sun, D. Yu, Controllable growth of well-aligned ZnO nanorod arrays by low-temperature wet chemical bath deposition method, Appl. Surf. Sci. 256 (2010) 1698–1702. https://doi.org/10.1016/j.apsusc.2009.09.097.

[21] L.L. Yang, Q.X. Zhao, M. Willander, Size-controlled growth of well-aligned ZnO nanorod arrays with two-step chemical bath deposition method, J. Alloys Compd. 469 (2009) 623–629. https://doi.org/10.1016/j.jallcom.2008.08.002.

[22] R. Parize, J. Garnier, O. Chaix-Pluchery, C. Verrier, E. Appert, V. Consonni, Effects of Hexamethylenetetramine on the Nucleation and Radial Growth of ZnO Nanowires by Chemical Bath Deposition, J. Phys. Chem. C. 120 (2016) 5242–5250. https://doi.org/10.1021/acs.jpcc.6b00479.

[23] M. Pelaez, N.T. Nolan, S.C. Pillai, M.K. Seery, P. Falaras, A.G. Kontos, P.S.M. Dunlop, J.W.J. Hamilton, J.A. Byrne, K. O'Shea, M.H. Entezari, D.D. Dionysiou, A review on the visible light active titanium dioxide photocatalysts for environmental applications, Appl. Catal. B Environ. 125 (2012) 331–349. https://doi.org/10.1016/j.apcatb.2012.05.036.

[24] R. Niyuki, H. Fujiwara, T. Nakamura, Y. Ishikawa, N. Koshizaki, T. Tsuji, K. Sasaki, Double threshold behavior in a resonance-controlled ZnO random laser, APL Photonics. 2 (2017). https://doi.org/10.1063/1.4974334.

Acknowledgement

This work was supported by Malaysia Ministry of Education through the Fundamental Research Grant Scheme (FRGS) under account number: 203.PFIZIK.6711667 and 2020 SATU Joint Research Scheme (JRS) NCKU 16.

Internet Connectivity: Is Free Space Optics the Solution?

SS Rais[1*], H Husni[2]

[1]School of Electrical Engineering, College of Engineering, University Teknologi Mara Cawangan Terengganu Kampus Dungun, 23000 Dungun, Malaysia
[2]School of Electrical Engineering, College of Engineering, Pusat Asasi UiTM, Universiti Teknologi MARA Malaysia Kampus Dengkil, 43800 Dengkil, Selangor

*Corresponding author's email: sitisara851@uitm.edu.my

Abstract. During the Coronavirus disease 2019 pandemic era, poor internet availability hampered online activity, particularly in the sector of education in Malaysia. The most effective way to address this situation is through the construction of telecommunications towers. However, it takes time due to the significant costs involved. Outdoor use of FSO appears to be more advantageous due to the absence of licensing requirements and the high bandwidth it offers. However, there are limitations. We will explore these FSO restrictions in this paper.

Keywords: Outdoor FSO; Internet access connectivity; Online Education

1. Introduction

Kosmo reported that the poor internet access in the rural areas of Gua Musang district in Kelantan is a significant challenge for Orang Asli students to follow Online Home-Based Learning (OHBL) [1]. They could not access the internet when they were at home and had to go out to the hilly areas of the village to access the internet [1]. The same goes for some students who had to stand outside their house to attend OHBL in Kuantan, Pahang, reported by Kosmo on April 29 2021[2]. On June 16 2021, the Borneo Post reported that a group of students from Kampung Entakong, Sarawak made part of the dirt road a classroom. They had to study on the dirt road due to poor internet access during OHBL [3]. Since the Covid-19 outbreak, we continue to hear about poor internet connections in some suburbs and rural places in Malaysia to this day.

As a result of poor internet connection, many things happen, such as difficulty in implementing OHBL by teachers, students are increasingly stressed, students cannot attend or complete assignments, and risks when engaging in OHBL. These situations have been reported by the news in [4-8].

The most effective way for internet connection is to build numerous additional telecommunications towers, as shown in Figure 1. If it is that simple, why has it not been done long ago? Even if these telecommunications towers are constructed, not all residents, especially the less fortunate, can afford a paid internet subscription [9]. According to Utusan Malaysia, students shared their mother's internet data for OHBL [10]. Malaysia Kini also stated that some families could not purchase internet data for their children's OHBLs due to budgetary restraints. For them, even purchasing basic necessities was already a challenge [11].

Figure 1. ISP network coverage.

Recognizing Malaysia's internet access challenges, the government should be more proactive in ensuring internet connectivity across the entire nation, be it urban or rural areas. Besides that, the private sectors through their Corporate Social Responsibility as well as individual contributors can support in providing internet access towards their local communities.

This article addressed how to implement internet connectivity via free space optical communication (FSO). Can it be implemented practically in Malaysia to address the issues raised? However, further options exist that are not covered in this study.

2. Internet Access Coverage

We present two (2) instances of circumstances with limited internet connectivity in this paper: (a) the Internet Service Providers (ISPs) coverage is limited, and (b) there is no fiber or cellular coverage, just satellite communications, as depicted in Figure 2. Throughout Malaysia, satellite broadband internet services are available [12]. However, it is not widely adopted due to the high cost involved. Both the end-user satellite devices and the internet plan are expensive for individual subscription [13]. In February 2021, the same ISP, in collaboration with the Malaysian government, presented another strategy in providing affordable internet plans [14]. Although the package offered is relatively affordable without paying the devices in rural areas of Sabah and Sarawak, the coverage is only limited to 100 meters from the transceiver station [15].

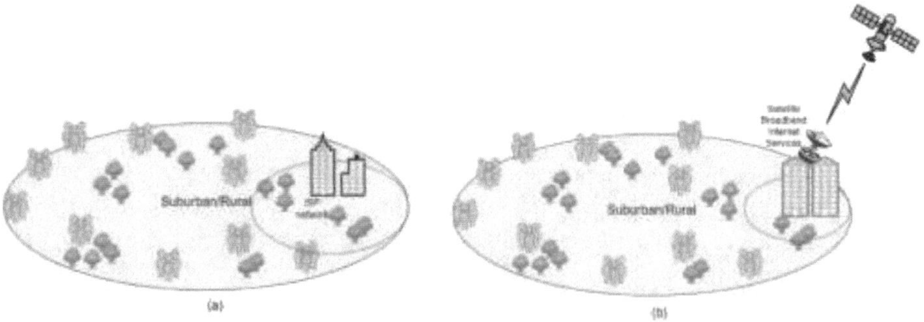

Figure 2. Limited connectivity in suburban/rural areas.

3. Internet Access Coverage Outdoor Free Space Optical (FSO) Communications

FSO is one of the branches of optical wireless communications (OWC) technologies [16]. Radio-based wireless communications have been widely used to address the rising demand for wireless capacity, but their scalability and bandwidth are limited [17]. Therefore, FSO technology is deployed to overcome these drawbacks. In addition, FSO has no spectrum licensing requirements [18].

Outdoor FSO networks provide long-distance point-to-point outdoor optical communications using the Line-of-Sight (LOS) mechanism in the atmosphere [19]. When a fiber-optic cable is unavailable or too expensive to build, outdoor FSO is an option for last-mile connectivity [18], as demonstrated in Figure 3. In the market today, optical wireless Point-to-Point connection distance up to 15 km [20]. Thus, through deployment of FSO, network coverage can be further expanded and benefit more people within that area.

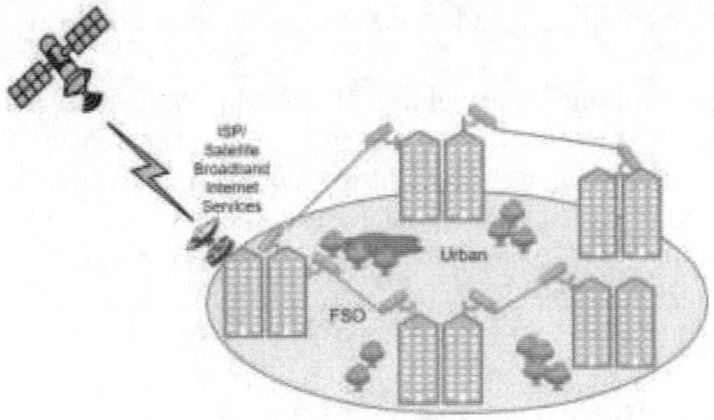

Figure 3. FSO links to scalable the network.

However, FSO has limitations based on the context as illustrated in Figure 2.

- Outdoor and last mile access: it can be deployed by service providers [21]. Thus, it does not differ from satellite communication systems. It will involve a higher cost, and there will be delay in its implementation.
- Building-to-building applications such as LAN-to-LAN, enterprise network or a campus network [20].
- Due to atmospheric loss, it needs a hybrid network design such as radio-based communication (RF) + FSO to provide optimum connectivity [17].

4. Conclusion

As it turns out, FSO is extremely valuable in the context of broadband internet services. However, several constraints are to be addressed in a situation such as that depicted in Figure 2. It entails ISPs, costs, and a relatively lengthy deployment phase. We recommend a few quick deployment techniques, but they are not discussed here. These techniques are our future works.

5. References

[1] A. T. Lah, "Akses Internet Lemah, Cabaran PdPR Pelajar Orang Asli," Kosmo!, Jan. 20, 2021.
[2] "Capaian Internet Ganggu PdPR," Kosmo!, Apr. 29, 2021.
[3] N. Syamimi, "Jaringan Internet Lemah, Pelajar Terpaksa Belajar Atas Tanah Untuk Sesi PdPR," The Borneo Post, Jun. 16, 2021.
[4] M. A. Jalil, "Capaian internet lemah sukarkan guru laksana PdPR," BH Online, Jan. 21, 2021.
[5] S. H. K. Arifin, "Calon SPM luar bandar tertekan PdPR dalam keadaan capaian internet lemah," Astro Awani, Jun. 08, 2021.
[6] H. L. Kamarudin, "Capaian internet lemah, pelajar terpaksa 'ponteng' PdPR," Sinar Harian, Jun. 28, 2021.
[7] "Pelajar parah lepas kemalangan ketika cari capaian internet untuk PdPR," Malaysia Kini, Jun. 20, 2021.
[8] "Pelajar Kemalangan Ketika Cari Capaian Internet Untuk PdPR," Utusan Borneo, Jun. 21, 2021.
[9] J. Sudrajat, "Kompetensi Guru Di Masa Pandemi Covid-19," J. Ris. Ekon. dan Bisnis, vol. 13, no. 2, p. 100, 2020, doi: 10.26623/jreb.v13i2.2434.
[10] J. Abdullah, "PdPR: Pelajar kongsi data internet ibu," Utusan Malaysia, Jun. 19, 2021.

[11] A. Lee and S. Buang, "Jangankan nak beli data, makan pun susah, keluh ibu," Malaysia Kini, 2020.
[12] "MEASAT FLEET," Measat Satellite Systems Sdn Bhd. http://www.measat.com/measat-fleet/ (accessed Jul. 07, 2021).
[13] A. Wong, "Connectme Satellite Broadband now costs a lot more to subscribe," Soyacincau.com, Jul. 18, 2020.
[14] "Satellite Broadband: An Immediate Solution To Digital Inclusion," CONNECTmeNOW, 2021. https://connectmenow.my/knowledge-center-post/19 (accessed Jul. 07, 2021).
[15] "How It Works," CONNECTmeNOW. https://connectmenow.my/#about (accessed Jul. 07, 2021).
[16] N. Balani, P. Kalamdar, D. Pillay, K. Ladekar, and K. Vishwakarma, "Review on Optical Wireless Communication Technologies," Int. J. All Res. Educ. Sci. Methods, vol. 9, no. May, pp. 1174–1177, 2021, doi: 10.13140/RG.2.2.28830.66887.
[17] I. K. Son and S. Mao, "A survey of free space optical networks," Digit. Commun. Networks, vol. 3, no. 2, pp. 67–77, 2017, doi: 10.1016/j.dcan.2016.11.002.
[18] S. A. Al-Gailani et al., "A Survey of Free Space Optics (FSO) Communication Systems, Links, and Networks," IEEE Access, vol. 9, pp. 7353–7373, 2021, doi: 10.1109/ACCESS.2020.3048049.
[19] A. Jahid, M. H. Alsharif, and T. J. Hall, "A Contemporary Survey on Free Space Optical Communication: Potential, Technical Challenges, Recent Advances and Research Direction," 2020, [Online]. Available: http://arxiv.org/abs/2012.00155.
[20] "Free Space Optics," EC System. http://www.ecsystem.cz/en/products/free-space-optic-equipment (accessed Jul. 07, 2021).
[21] A. Malik and P. Singh, "Free Space Optics: Current Applications and Future Challenges," Int. J. Opt., vol. 2015, no. c, 2015, doi: 10.1155/2015/945483.

A Study of Optical Micro-bottle Resonator Humidity Sensor

Ummu Umairah Mohamad Ali[1*], Ahmad Ashadi Johari[2], Zulzilawati Jusoh[3*], Husna Abdul Rahman[1], Sulaiman Wadi Harun[4]

[1]School of Electrical Engineering, College of Engineering, University Teknologi Mara, Shah Alam 40450, Malaysia
[2]Universiti Teknikal Malaysia Melaka (UTeM), Hang Tuah Jaya, 76100 Durian Tunggal, Melaka, Malaysia
[3]School of Electrical Engineering, College of Engineering, University Teknologi Mara Cawangan Terengganu Kampus Dungun, 23000 Dungun, Malaysia
[4] Photonics Engineering Laboratory, Department of Electrical Engineering, University of Malaya, Kuala Lumpur 50603, Malaysia

*Corresponding author's email: umairah.ali@gmail.com, zulzilawati@uitm.edu.my

Abstract. This experiment reports the fabrication of the optical Micro-bottle Resonator (MBR) as a relative humidity (RH) sensor. The MBR is fabricated via the "soften-and-compress" method to create the bottle-shaped with diameter Db = 175 um. The MBR is then optically excited using a 8 μm microfiber and was found to have a Q Factor of $>10^5$. The MBR was then employed as a humidity sensor with the RH level percentage range between 40% to 100% and the performance is compared with a non MBR microfiber. The MBR sensor was found to have a sensitivity of 0.2973 dBm/%, linearity >90% and and is superior to the bare microfiber in all measured parameters . The MBR humidity sensor was also found to have good repeatability and stability over a period of 60 seconds compare to bare microfiber.

1. Introduction

Recently, the micro-resonator (MR) has received lot of attention due to their wide applications. By supporting the whispering gallery mode (WGM), OMRs have gained a potential towards application in optical microsystems and miniaturization attention [1,2]. They're usually associated with circular-path resonant cavities like micro-sphere, micro-ring and micro-disc geometry and these resonators can support whispering gallery mode (WGM) operation by providing continuous internal reflection at specified resonant wavelengths [3-12]. A WGM can be represented as an optical beam transmitted solely at the micro-resonator surface due to grazing-angle total internal reflection, according to the geometrical optical principle [13]. When light is evanescently coupled into the micro-resonator via a microfiber, the transmission spectrum shows a very tiny resonance dip with a full width half maximum (FWHM) on a level of pm. Optical sensing is one of the many applications for the optical WGM device [14-15].

Relative humidity (RH) sensors based on optical fiber have been extensively explored for various important application such as agriculture and in medical field [16]. RH sensors previously have been described using a variety of approach including tapered fiber, fiber grating, coupler microfiber, interferometer fiber and resonator based humidity sensor [16]. In this studies, we study the characteristics of micro-bottle resonator (MBR) coupled with microfiber with size 8 μm on the sensing performance. The MBR formed with technique known as "soften-and-compress" on SMF-28 silica fiber using manual splicer machine while microfiber was fabricating by tapering machine. Thinner microfibers may improve light coupling to WGM resonators because the fiber was physically in contact with them. The position of MBR were in the middle area of taper microfiber where the distance is always touch each other respectively. They were then used to measure humidity in a range 40% to 100%. Analysis of the performance of MBR coupled tapered exposed the potential of MBR with tapered bare fiber in humidity sensing.

2. Experiment Setup

MBR that are using in this study were made with the so called "soften-and-compress" method [2] using a splicer machine (Furukawa Electric Fitel S178A) as shown in **Figure 1**. In a mode manual splicer machine, placed some length of SMF-28 fiber and clamped on both sides and a small center part is heated under a plasma arc. This clamped fiber was compressed inward in the direction of the plasma arc, forming bottle like structure.

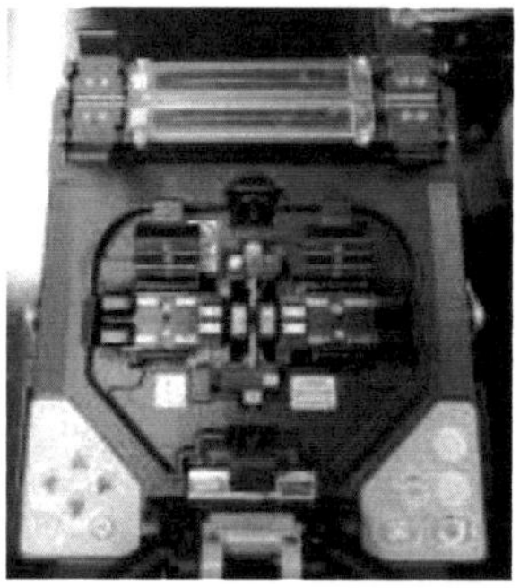

Figure 1. A splicer machine used to form micro-bottle resonator

The size is determined by the arc number during the fabrication process. **Figure 2** shows microscopic image of the of MBR resonator whereas with 10 time's arc. The size a MBR is defined by diameter stem, Ds and the bottle diameter, Db. MBR in this work has stem, Ds and bottle diameter, Db of 125 μm and 175 μm, as shown in **Figure 2**.

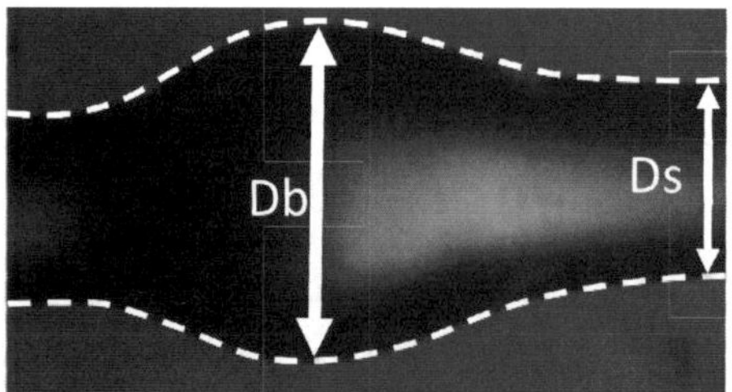

Figure 2. The micro-bottle resonator (MBR)

To allow for coupling of light to and from the MBR, a microfiber was fabricated using a customized flame brushing technique with 8 μm tapering waist. **Figure 3** shows the MBR, which was placed directly on the microfiber to allow efficient coupling of light in or out of the MBR. This is attributed to the evanescent field of a phase-matched microfiber can be easily aligned to be overlap with the evanescent field of a WGM from the MBR.

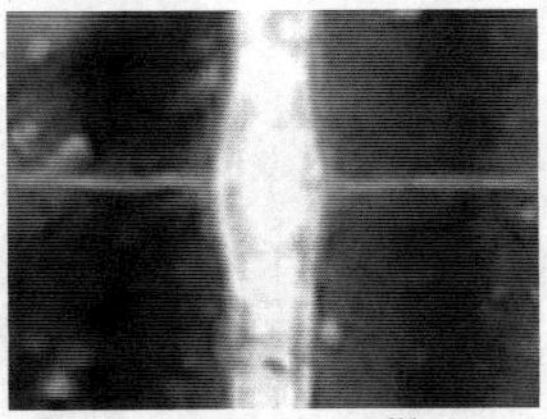

Figure 3. Tapered microfiber and MBR

Figure 4 shows the schematic diagram of the experiment humidity sensor uses an MBR as sensor. Tunable laser source (TLS) (ANDO AQ4321D) is a source the light to the MBR structure via a microfiber and the output of the MBR was detected using an optical power meter (OPM) (THORLABS PM100D). The MBR structure was placed in a control chamber, which percentage humidity level varies from 40%-100% and the humidity controlled by saturated salts solution inside the chamber. Then, the relative humidity was recorded by a humidity meter (Hygrometer RS 1365, Sensitivity: 1%) and sensing media's output power of the sensing was continually measured. The OPM was also connected to the computer, making it easier to display the result and generate the graph.

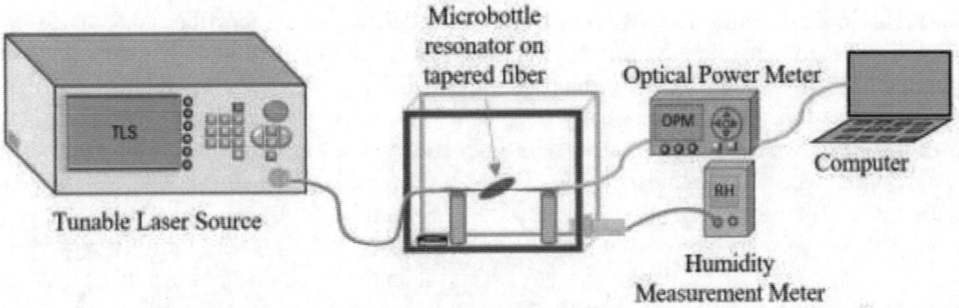

Figure 4. Humidity sensor experimental setup

3. Result and Discussion

The MBR was first characterized using TLS together with an OPM and the TLS, which its operations in the wavelength range between 1520 nm and 1620 nm and average output power of 1 dB has been launched into the microfiber coupled MBR. OPM was used to measure the output power of a laser that adjusted wavelength between from 1520.2 nm to 1520.3 nm with a wavelength interval of 0.001 nm. From the **Figure 5** below, the Q factor for MBR is that coupled with microfiber is 7.6013×10^5 dBm. The Q-factor loss can be influenced by altering the position or spacing between MBR with tapered microfiber. The insertion loss of MBR is roughly 27.5 dBm to 31 dBm. The MBR's Q-factor, defined as $\Delta\lambda/\lambda$ where λ is the resonant wavelength. This is thought to be due to the microfiber, which contributes significantly to the entire micro-bottle ensemble's insertion loss.

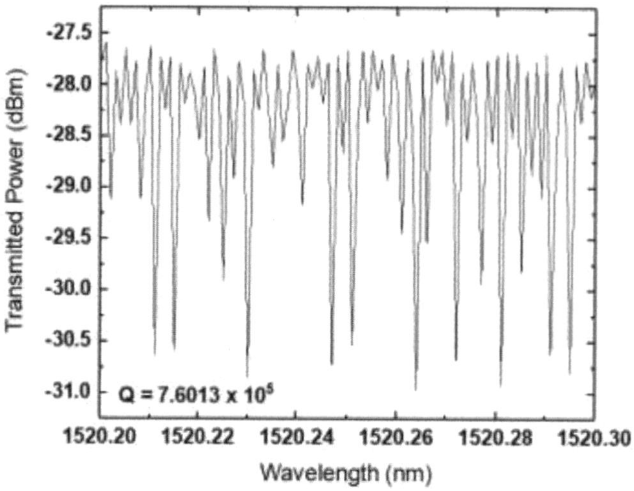

Figure 5. Transmitted Power of MBR varies with wavelength

For bare microfiber and MBR, the fluctuation of transmitted light against relative humidity was examined. Figure 6 depicts the fluctuation of transmitted light in bare microfiber and MBR on tapered microfiber as a function of relative humidity. As the relative humidity rises from 40% to 100%, the intensity of the transmitted light through the bare micro-fiber and MBR on the tapered fiber increases. **Table 1** summarizes the performance characteristics of raw microfiber and MBR. The MBR coupled with microfiber is shown much better than bare microfiber. The sensitivity of the bare microfiber is 0.152 dB/%, 38.99 % of a slope linearity and 20.197 % of a limit of detection, whereas the sensitivity of the tapered microfiber and MBR is 0.2973 dB/%, 97.80% of a slope linearity and 19.132 % of limit of detection. The analysis discovered that the MBR has a substantially higher humidity sensing efficacy than bare microfiber. One of the cause is surface absorption in the resonator's transmitting insertion loss, which increases as humidity levels. Despite the fact that losses in the resonator were observed as a result of light circulation inside the resonators, power loss with each circle pathway considerably increased the sensitivity.

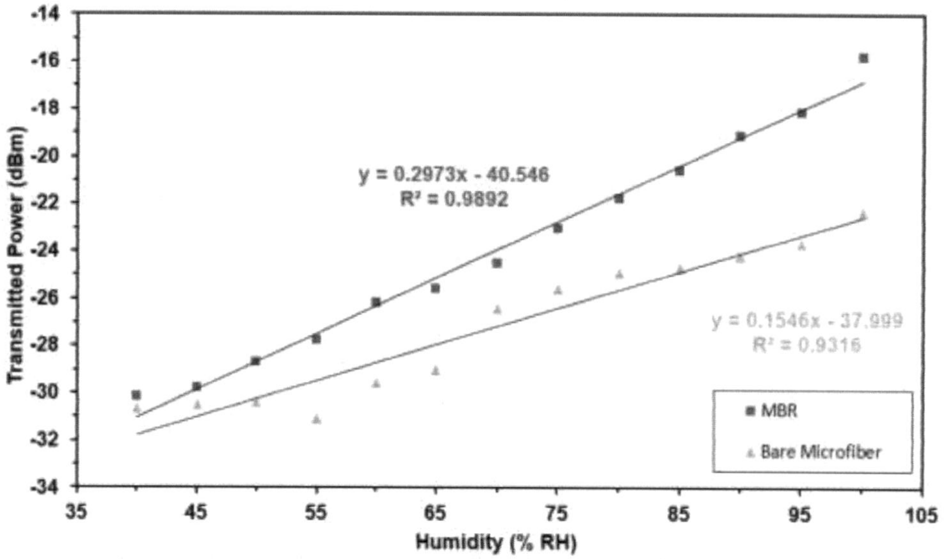

Figure 6. The bare microfiber and MBR response toward humidity sensing

Table 1. SENSING PERFORMANCE OF HUMIDITY SENSING.

Parameters	Bare microfiber	MBR
Linearity (%)	38.99%	97.80%
Sensitivity (dBm/%RH)	0.1546	0.2973
Standard deviation (dBm)	3.00	4.46
Linear Range Humidity (%)	40-100	40 - 100
Limit of detection	20.197	19.132

Every RH experiment was repeated three times for MBR and bare microfiber to determine the setup's reproducibility. **Figure 7** shows that the results were similar for MBR and bare microfiber, with MBR sensitivity values of more than 0.25 dB/% and bare microfiber sensitivity value of less than 0.2 dB/%, respectively. %. As a result, when compared to the bare microfiber, the MBR functioned far better as a humidity sensor.

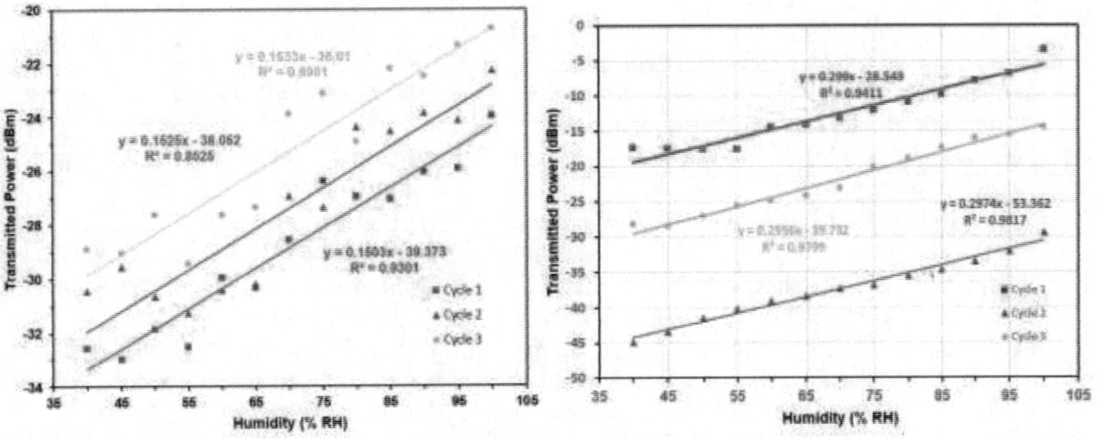

(a) (b)

Figure 7. The MBR RH sensing repeatability performance of (a) bare microfiber and (b) MBR

4. Conclusion

This paper presented the performance of MBR based humidity sensor using micro-fiber. A micro-bottle resonator (MBR) has a stem and bottle diameter of 125 μm, and 175 μm, respectively. MBR was made using a 'soften-and-compressed' method was applied to a silica microfiber which formed bulge area at the center of fiber. The MBR was then excited through the microfibers with the diameters of 8 μm via a TLS. With increase the RH level from 40% to 100%, the MBR was employed as a humidity sensor. The MBR RH sensor was observed to have a sensitivity of 0.2973 dB/%, linearity >90% and showed relatively good repeatability and stability. This simple humidity sensor can offer wide range of applications such as in green house automatic control system and in extremely humid environments.

References

[1] K.J. Vahala, Optical microcavities, *Nature* 424 (6950) (2003) 839.
[2] A.B. Matsko, V.S. Ilchenko, Optical resonators with whispering gallery modes I: basics, *IEEE J. Sel. Top. Quantum Electron* 12 (3) (2006) 3.
[3] Sumetsky, M., Dulashko, Y., & Windeler, R. (2010). Optical microbubble resonator. *Optics Letters, 35*(7), 898-900.
[4] Chiasera, A., et al., *Spherical whispering-gallery-mode microresonators.* Laser & Photonics Reviews, 2010. **4**(3): p. 457-482.
[5] Murugan, G.S., J.S. Wilkinson, and M.N. Zervas, *Selective excitationof whispering gallery-modes in a novel bottle microresonators.* Optica express, 2009. **17**(14): p. 11916-11925.
[6] Little, B.E., et al., *Microring resonator channel dropping filters.* Journal of lightwave technology, 1997. **15**(6): p. 998-1005.
[7] Absil, P.P., *Microring resonators for wavelength division multiplexing and integrated photonics applications.* 2000.
[8] Kokubun, Y., *Wavelength selective integrated device by vertically coupled microring resonator filter.* Photonics Based on Wavelength Integration and Manipulation IPAP Books, 2005. **2**: p. 303-316.
[9] Armani, D., et al., *Ultra-high-Q toroid microcavity on a chip.* Nature, 2003. **421**(6926): p. 925.
[10] Armani, D., et al., *Electrical thermo-optic tuning of ultrahigh-Q microtoroid resonators.* Applied physics letters, 2004. **85**(22): p. 5439-5441.
[11] Gérard, J., et al., *Enhanced spontaneous emission by quantum boxes in a monolithic optical microcavity.* Physical review letters, 1998. **81**(5): p. 1110.
[12] Sarma, J. and K. Shore, *Electromagnetic theory for optical disc resonators.* IEE Proceedings J (Optoelectronics), 1985. **132**(6): p. 325-330.
[13] Sumetsky, M., & Fini, J. (2011). Surface nanoscale axial photonics. *Optics express, 19*(27), 26470-26485.
[14] Marcatili, E.A.J., "*Bends in optical dielectric waveguides*", Bell System Techn. J. 48, 2103-2132 (1969).
[15] Braginsky, V.B., Gorodetsky, M.L., and Ilchenko, V.S. *Quality-factor and nonlinear properties of optical whispering-gallery modes.* Phys. Lett. A 137, 393-396 (1989).
[16] Lee, P.-T., et al., *High quality factor microcavity lasers realized by circular photonic crystal with isotropic photonic band gap effect.* Applied physics letters, 2007. **90**(15): p. 151125.
[17] Bianucci, P., *Optical microbottle resonators for sensing.* Sensors, 2016. **16**(11): p. 1841.
[18] Y. Peng, Y. Zhao, M.Q. Chen, F.Xia, *Research advances in microfiber humidity sensors*. Small 14 (29) (2018) 1-20

Acknowledgement

The authors would like to thank the Photonics Engineering Laboratory, University of Malaya, for the research facility provided throughout this work, and also thank to Universiti Teknologi MARA for financially support this research work.

Hafnium Bismuth Erbium Co-Doped Fiber Based Dark Pulses Generation With Black Phosphorus As Saturable Absorber

A Ahmad[1*], M F A Rahman[1], M A M Johari[1], A A Latiff[2], M H Jali[3], H H M Yusof[3], X S Cheng[4], A R Muhammad[5], and S W Harun[6]

[1]Fakulti Teknologi Kejuruteraan Elektrikal dan Elektronik, Universiti Teknikal Malaysia Melaka, Hang Tuah Jaya,76100 Durian Tunggal, Melaka, Malaysia
[2]Fakulti Kejuruteraan Elektronik dan Kejuruteraan Komputer, Universiti Teknikal Malaysia Melaka, Hang Tuah Jaya, 76100 Durian Tunggal, Melaka, Malaysia
[3]Fakulti Kejuruteraan Elektrik, Universiti Teknikal Malaysia Melaka, Hang Tuah Jaya, 76100 Durian Tunggal, Melaka, Malaysia
[4]School of Engineering, KDU University College, Utropolis, Glenmarie Jalan Kontraktor U1/14, Shah Alam, Selangor Darul Ehsan, Malaysia.
[5]Institute of Microengineering and Nanoelectronics (IMEN), University Kebangsaan Malaysia, 43600 UKM, Bangi Selangor, Malaysia
[6]Department of Electrical Engineering, University of Malaya, 50603, Kuala Lumpur, Malaysia

[*]Corresponding author's email: aminah@utem.edu.my

Abstract A dark pulse generation is demonstrated in a fiber laser configured with a 20 cm long HBEDF and multilayer Black Phosphorus as a gain medium and saturable absorber, respectively. Dark pulses fiber laser at 1.5 µm region was obtained when the pump power exceeds the threshold of 147 mW. Meanwhile, the spectrum of the dark pulse is centred at 1556.40 nm, with the 3 dB bandwidth of 0.12 nm and the separation between adjacent pulses is 1145 ns, corresponding to the cavity length of 211 m. The pulse width was measured to be around 320 ns. The radio-frequency spectrum of the dark pulse, which was measured within the 20 MHz range. More than 17 harmonics were observed within this range, which indicates the mode-locking operation of the laser. The fundamental frequency was obtained at 1.1 MHz, which agreed with the oscilloscope trace. Furthermore, it shows a signal to noise ratio of about 36.58 dB, which indicates good stability. The maximum output power of 0.78 mW and pulse energy of 0.78 nJ were obtained at 187 mW pump power.
Keywords: Fiber lasers, Dark pulsed laser, Mode-locking.

1. Introduction
The advantages of compactness and flexibility of fiber lasers have drawn widespread attention. Most of the applications used is being practised in many applications such as in optical communication, fiber sensor technologies, micromachining and the military system [1-3]. Q-switching and mode locking are mostly used by these applications. Compared to the conventional active technique, passive mode-locked is desirable through the usage of a saturable absorber (SA) which produce an ultrashort pulsed fibre laser[4]. This is due to the rapid modulation of the resonator compared to any electronic modulator which is necessary for an actively mode-locked laser [4]. Recently, numerous types of saturated absorbing materials are used for pulse laser generation, including semiconductor saturable absorber mirror (SESAM) [5], single-walled carbon nanotubes (SWCNTs) [6] and graphene [7, 8].

Most of the researchers have used semiconductor saturable absorber mirrors (SESAMs) for passively generating mode locked lasers even though their fabrication process is complicated and expensive [9]. In addition, the operating wavelength depends on the semiconductor materials. Nowadays popular passive techniques are SWCNTs and graphene. However, the SWCNTs SA operation wavelength is determined by the nanotube diameter and the bandgap engineering was complicated, which normally lead to an uncontrollable non-saturable loss. Meanwhile, graphene has higher electron mobility and zeroes energy bandgap, which allows broadband operation. However, its optical absorption is located weakly at 1.5 μm region, so its application in the optical communication area may be limited. Apart from SWCNTs and graphene, many other nanomaterials have been investigated in fiber laser systems as SA devices, such as topological insulators, transition metal dichalcogenides and black phosphorus (BP). BP have a narrow direct band-gap and wide optical response from infrared to mid-infrared that corresponds to the bulk of layers and single, respectively[4].

There is also interest in the generation of dark pulse lasers that are less sensitive to fiber loss compared to bright pulse [9, 10]. Dark Pulse laser emits a steady beam of light, with periodic dips in the continuous bright background. This type of lasers are new, so their applications are still under investigation [11]. Optical frequency combs and optical atomic clocks are some of the applications for dark pulse lasers. The frequency of dark pulse waves could be used to transfer information. Meanwhile, the continuous-wave background may provide a single strong comb line that could be used to probe a quantum transition. In telecommunications, some applications can be used due to the lack of dispersion and linearity during the transmission of dark pulses [12].

Many works have been previously reported on the generation of dark pulses. For instance, Wang et al. demonstrated the generation of dark pulses in a mode-locked Erbium-doped fiber laser (EDFL) based on a Molybdenum disulfide film based saturable absorber (SA) where the laser was at a 1.7 MHz repetition rate[13]. In another work, Zhao et al. produced dark pulses by using rhenium disulfide (ReS_2) as a saturable absorber [14].

Mode-locked fiber lasers operating at 1.5 μm region were demonstrated using 20 cm long Hafnium Bismuth Erbium Doped Fiber (HBEDF) as a gain medium in conjunction with the newly developed passive saturable absorber. The HBEDF was achieved by the combination of Hf, Bi, and Er doped yttria alumina-silica glass based preform, which have absorption loss at 980 nm is found to be 100 dB/m equivalent to 12 500 wt ppm [15, 16] Here, the pulse generation was realised using Black Phosphorus as a saturable absorber.

2. Methodology

2.1 Pulse generation with BP SA

In 2D nanomaterials, BP has gained much interest in recent years as a potential in many applications. The multilayer BP has good optical characteristics such as wideband absorption and ultrafast carrier dynamics. Furthermore, it comprises only the elemental "phosphorus", and thus it could be easily peeled off by mechanical exfoliation. The BP SA was prepared by transferring multilayer BP onto a fiber ferrule tip using a mechanical exfoliation method. At first, thin flakes were moderately peeled off from a big block of commercially available BP crystal (purity of 99.995 %) using clear scotch tape. Then, the flakes were repeatedly pressed so that they adhered onto the scotch tape to form a thin layer of BP. Afterwards, an end surface of fresh standard FC/PC fiber ferrule tip was pressed down on the scotch tape to transfer the multilayer BP onto it. The BP transferring process is described in Figure 1 (a). Next, the ferrule with multilayer BP was connected to another fresh FC/PC fiber ferrule via a fiber adapter to form an all-fiber BP based SA device. A tiny amount of index matching gel was applied at the connector to minimise the SA device loss. The energy dispersive spectroscopy (EDS) analysis was used to examine the composition of the multilayer BP tape[17]. The presence of BP material on the scotch tape adhesive surface was confirmed by the presence of the high peak of phosphorus in the spectroscopy, as shown in Figure 1 (b). Figure 1 (c) shows the FESEM image of the multilayer BP tape, which confirmed the existence of uniform multilayer phosphorus on the tape.

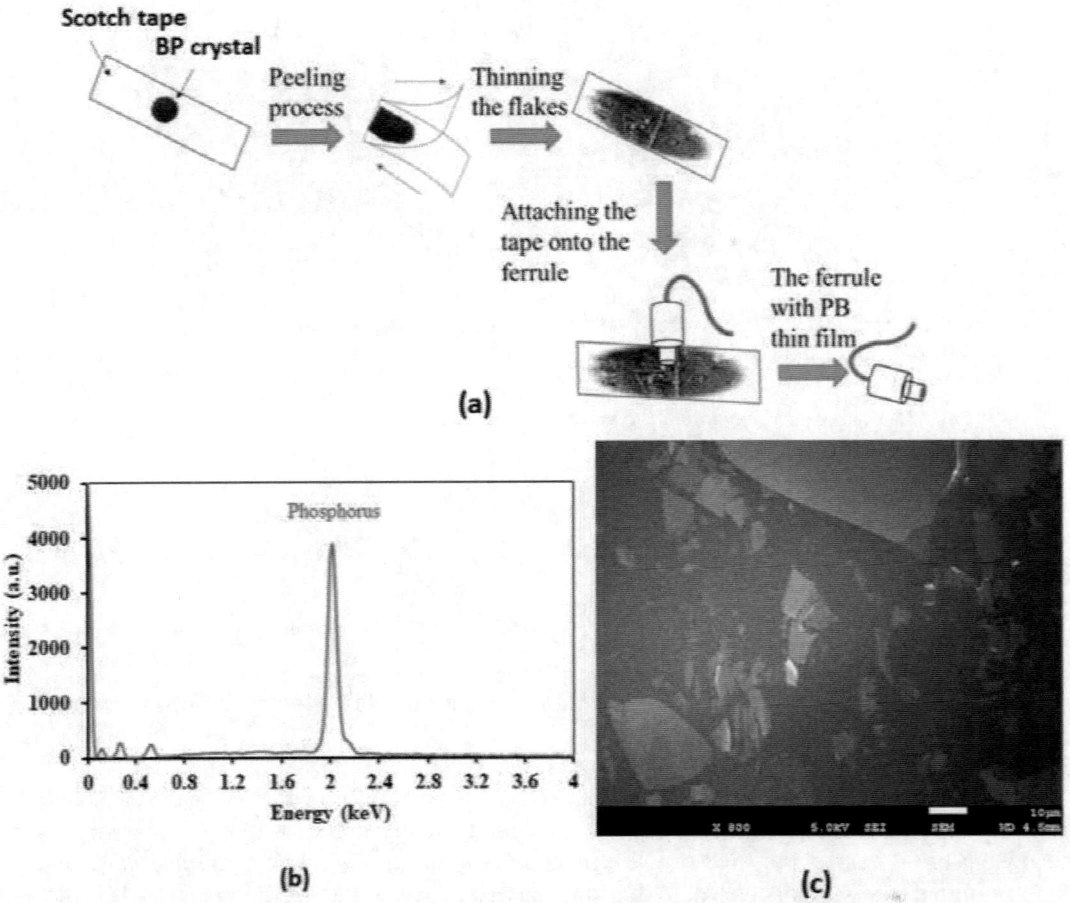

Figure 1. Mechanical exfoliation of the multilayer BP (a) The BP transferring process (b) EDS analysis confirming the presence of phosphorous (c) FESEM image.

2.2 Laser Cavity Configuration

The laser cavity has a typical ring configuration, as shown in Figure 2. It uses a passive BP SA for mode-locking. A 20 cm HBEDF piece was pumped with a 980 nm laser diode via 980nm/1550nm wavelength division multiplexer (WDM). It produced an amplified spontaneous emission (ASE) photons, which oscillated in the laser cavity to produce lasing at 1550 nm region. The SA functions as a mode-locker to convert the continuous-wave lasing to nanosecond pulses. A polarisation insensitive optical isolator was incorporated inside the ring cavity to lock the propagation of light in one direction and thus to prevent any detrimental effects due to spurious reflections inside the resonator. A 198 m long single mode fiber (SMF) coupler was also inserted to increase the nonlinearity so that enough phase shift per round trip can be achieved in the cavity for assisting the mode-locking operation. A 10 dB output coupler was used to split the output power in the portion of 90% and 10%. The 90% portion was channelled back into the cavity for further oscillation, while 10% portion was tapped out as the output.

The spectral and temporal analysis of the Q-switched EDFL was carried out using a 0.02 nm resolution OSA (Yokogawa AQ6370C) and a high-speed photodetector linked to an oscilloscope (GWINSTEK: GDS-3352), respectively. A 7.8 GHz Radio Frequency (RF) spectrum analyser (Anritsu MS2683A) was used to measure the repetition rate and evaluate the stability of the pulse laser. The average laser power was measured by an optical power meter (Thorlabs PM 100D) coupled with an InGaAs powerhead operating between 800-1700 nm (Photodiode Power Sensor S145C Integrating Sphere).

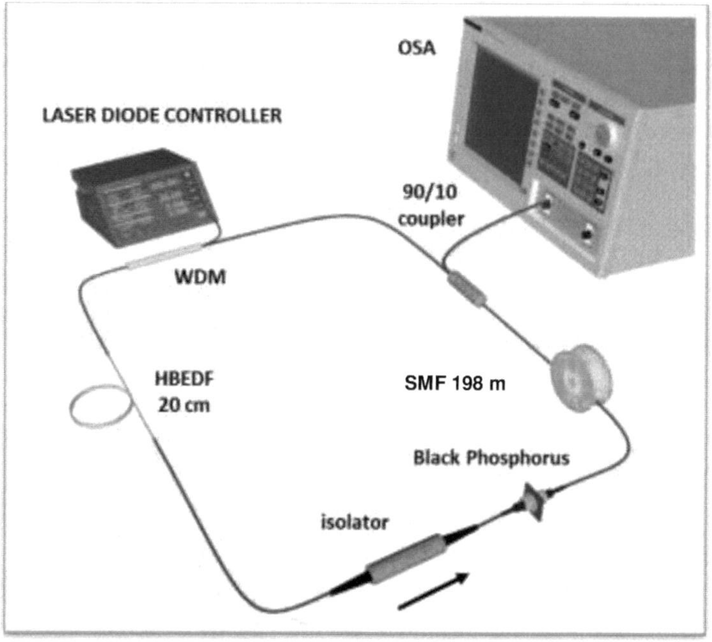

Figure 2. Experimental setup of the mode-locked fiber laser for dark pulses generation

3. Results and Discussion

In this experiment, the dark pulses were obtained when the pump power exceeds the threshold of 147 mW. As shown in Figure 3, the spectrum of the dark pulse is centred at 1556.40 nm, with the 3 dB bandwidth of 0.12 nm. Compared to CW operation (without the SA), the operating wavelength was slightly shifted to a shorter wavelength due to the insertion loss of the SA. Figure 4 displays the mode-locked pulse train, which indicates obvious dark pulses. The separation between adjacent pulses is 1145 ns, corresponding to the cavity length of 211 m. The pulse width was measured to be around 320 ns. The formation of dark pulses is most probably due to the deployment of highly nonlinear HBEDF in a long cavity. The cavity design provides an independent polarisation nature. Thus, the two orthogonal polarisation components inevitably exist and couple to each other, inducing the generation of the dark pulses. Figure 5 illustrates the radio-frequency spectrum of the dark pulse, which was measured within the 20 MHz range. More than 17 harmonics were observed within this range, which indicates the mode-locking operation of the laser. The fundamental frequency was obtained at 1.1 MHz, which agreed with the oscilloscope trace. It shows a signal to noise ratio of about 36.58 dB, which indicates good stability.

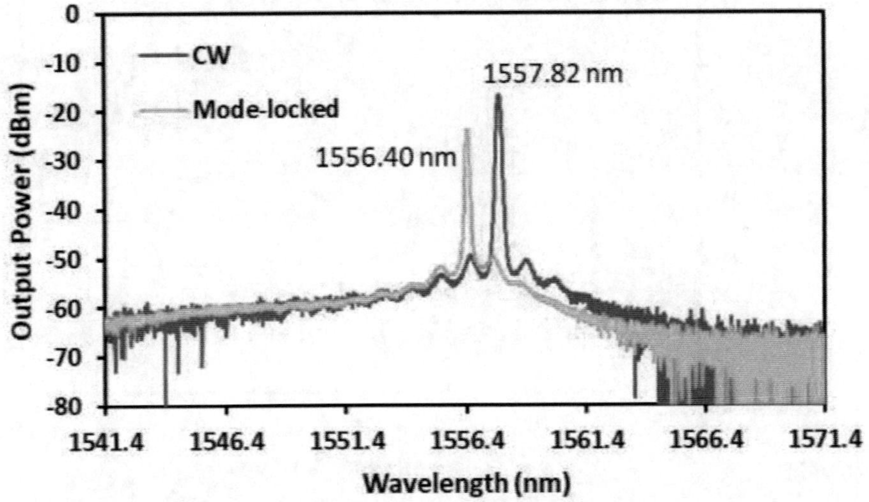

Figure 3. Output spectra of the fiber laser with and without SA at a pump power of 147 mW.

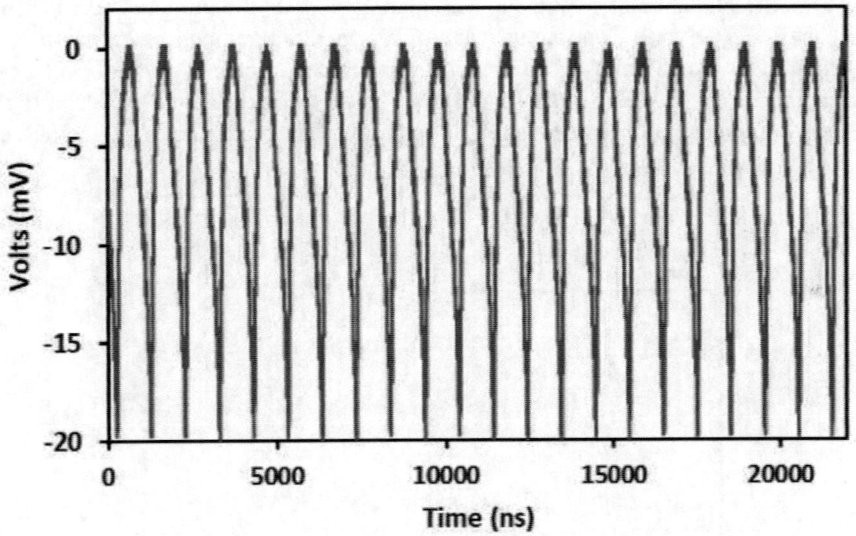

Figure 4. Typical oscilloscope trace for the dark pulses

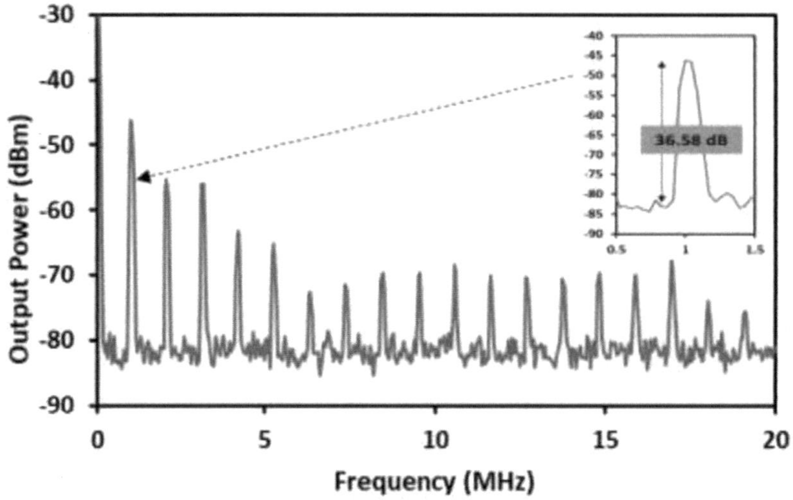

Figure 5. RF spectrum of the dark pulses

Figure 6. shows the output laser power and pulse energy versus the pump power. When the pump power is below the 147 mW, only continuous wave (CW) operation happens. Continuously increasing the pump power from 147 to 187 mW, a stable mode-locking operation is observed. Both output power and pulse energy increase with the rise of pump power. The maximum output power of 0.78 mW and pulse energy of 0.78 nJ were obtained at 187 mW pump power. The mode-locking operation ceases with a further increase in pump power above 187 mW.

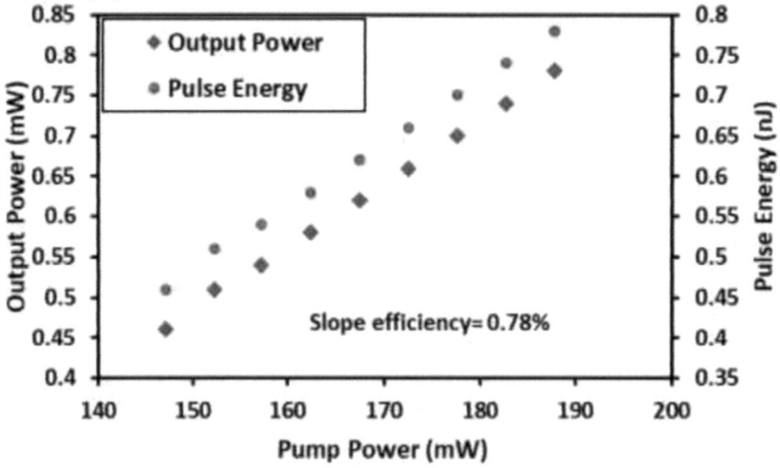

Figure 6. Output power and pulse energy of the mode-locked laser against pump power

4. Conclusion

This research demonstrates a dark pulse generation in a fiber laser configured with a 20 cm long HBEDF and Black Phosphorus as a gain medium and saturable absorber, respectively. The dark pulses fiber laser at 1.5 μm region was obtained when the pump power exceeds the threshold of 147 mW. The spectrum of the dark pulse is centred at 1556.40 nm, with the 3 dB bandwidth of 0.12 nm. The separation between adjacent pulses is 1145 ns, corresponding to the cavity length of 211 m. The pulse width was measured to be around 320 ns. The radio-frequency spectrum of the dark pulse, which was measured within the 20 MHz range. The maximum output power of 0.78 mW and pulse energy of 0.78 nJ were obtained at 187 mW pump power. Dark pulse lasers are new, so applications for them are still a matter of speculation. Some of the possible applications, for instance, are related to optical frequency combs and optical atomic clocks.

5. References

[1] M. A. M. Johari, M. H. B. Jali, H. H. B. M. Yusof, H. R. B. A. Rahim, A. B. Ahmad, M. I. M. A. Khudus, *et al.*, "Polyvinyl alcohol coating microbottle resonator on whispering gallery modes for ethanol liquid sensor," *Optics & Laser Technology,* vol. 143, p. 107379, 2021.

[2] M. H. Jali, H. R. A. Rahim, M. A. M. Johari, A. Ahmad, H. H. M. Yusof, S. H. Johari, *et al.*, "Integrating microsphere resonator and ZnO nanorods coated glass for humidity sensing application," *Optics & Laser Technology,* vol. 143, p. 107356, 2021.

[3] M. B. Hisyam, M. F. M. Rusdi, A. A. Latiff, and S. W. Harun, "Generation of mode-locked ytterbium doped fiber ring laser using few-layer black phosphorus as a saturable absorber," *IEEE Journal of Selected Topics in Quantum Electronics,* vol. 23, pp. 39-43, 2016.

[4] A. Markom, S. Tan, A. Muhammad, M. C. Paul, A. Dhar, S. Das, *et al.*, "Dark pulse mode-locked fibre laser with zirconia-based erbium-doped fibre (Zr-EDF) and Black phosphorus saturable absorber," *Optik,* vol. 223, p. 165635, 2020.

[5] U. Keller, K. J. Weingarten, F. X. Kartner, D. Kopf, B. Braun, I. D. Jung, *et al.*, "Semiconductor saturable absorber mirrors (SESAM's) for femtosecond to nanosecond pulse generation in solid-state lasers," *IEEE Journal of Selected Topics in Quantum Electronics,* vol. 2, pp. 435-453, 1996.

[6] S. W. Harun, M. A. Ismail, F. Ahmad, M. F. Ismail, R. M. Nor, N. R. Zulkepely, *et al.*, "A Q-Switched Erbium-Doped Fiber Laser with a Carbon Nanotube Based Saturable Absorber," *Chinese Physics Letters,* vol. 29, p. 114202, 2012.

[7] H.-P. Li, H.-D. Xia, Z.-G. Wang, X.-X. Zhang, Y.-F. Chen, S.-J. Zhang, *et al.*, "A compact graphene Q-switched erbium-doped fiber laser using optical circulator and tunable fiber Bragg grating," *Chinese Physics B,* vol. 23, p. 024209, 2014.

[8] J. Zhao, Y. Wang, P. Yan, S. Ruan, Y. Tsang, G. Zhang, *et al.*, "An Ytterbium-doped fiber laser with dark and Q-switched pulse generation using graphene-oxide as saturable absorber," *Optics Communications,* vol. 312, pp. 227-232, 2014/02/01/ 2014.

[9] B. Nizamani, S. Salam, A. Jafry, N. Zahir, N. Jurami, M. A. Khudus, *et al.*, "Indium tin oxide coated D-shape fiber as a saturable absorber for generating a dark pulse mode-locked laser," *Chinese Physics Letters,* vol. 37, p. 054202, 2020.

[10] Z. C. Tiu, S. J. Tan, H. Ahmad, and S. W. J. C. O. L. Harun, "Dark pulse emission in nonlinear polarization rotation-based multiwavelength mode-locked erbium-doped fiber laser," vol. 12, p. 113202, 2014.

[11] Y. S. Kivshar and B. J. P. r. Luther-Davies, "Dark optical solitons: physics and applications," vol. 298, pp. 81-197, 1998.

[12] J. Yu, L. Xu, Y.-K. Yeo, P. N. Ji, T. Wang, and G.-K. Chang, "A novel scheme for generating optical dark return-to-zero pulses and its application in a label switching optical network," *IEEE photonics technology letters,* vol. 18, pp. 1524-1526, 2006.

[13] R. Wang, Y. Yao, Q. Wu, Y. Yang, J. Tian, and K. Xu, "Bright-dark pulses produced by passively mode-locked fiber laser with Molybdenum disulfide saturable absorber," in *Advanced Laser Technology and Applications*, 2018, p. 108440J.

[14] R. Zhao, G. Li, B. Zhang, and J. J. O. e. He, "Multi-wavelength bright-dark pulse pair fiber laser based on rhenium disulfide," vol. 26, pp. 5819-5826, 2018.

[15] A. Ahmad, X. S. Cheng, M. C. Paul, A. Dhar, S. Das, H. Ahmad, *et al.*, "Investigation of the Brillouin effect in highly nonlinear hafnium bismuth erbium doped fiber," *Microwave and Optical Technology Letters,* vol. 61, pp. 173-177, 2019.

[16] A. Ahmad, S. W. Harun, M. C. Paul, M. Rusdi, S. Das, A. Dhar, *et al.*, "Bismuth-doped fiber as Q-switcher in hafnium bismuth erbium co-doped fiber laser," *Microwave and Optical Technology Letters,* vol. 62, pp. 3634-3639, 2020.
[17] T. A. Alghamdi, S. Adwan, H. Arof, and S. W. Harun, "Application of black phosphorus for pulse generation in erbium-doped fiber laser," *Results in Optics,* vol. 4, p. 100091, 2021.

Acknowledgements
The authors are grateful to Universiti Teknikal Malaysia Melaka for the research and financial support. The work is funded by Research Development Grant Scheme RAGS/1/2014/SG04/UTEM/2.

Whispering gallery mode resonance excitations on a partially gold coated bottle microresonator

Mohd Narizee Mohd Nasir[1,2*], **G S Murugan**[2] **and M N Zervas**[2]

[1] Jabatan Fizik Gunaan, Fakulti Sains dan Teknologi, Universiti Kebangsaan Malaysia, 43600, Bangi, Selangor
[2] Optoelectronics Research Centre, University of Southampton, Southampton, SO17 1BJ, United Kingdom

[*]Corresponding author's email: narizee@ukm.edu.my

Abstract. The study on the excitations of whispering gallery mode (WGM) resonances optical bottle microresonator (BMR) partially coated with a thin gold Au metal film is presented. The BMR was fabricated through "soften-and-compress" technique on a small section of standard optical fibre. Depositing Au particles on the spheroidal curvature of the BMR surface yields a thin metal-film of a meniscus profile with 200 nm maximum gold thickness and tapered edges. A polarization resolved experimental setup was used to excite TE- and TM-mode resonances. Coupling strengths of the excited WGMs would vary with different coupling arrangements relative to the position of the meniscus Au film. Calculated *Q*-factor values of composite TE and TM mode resonances were determined to be in the range 1800 and 2700, respectively.

1. Introduction

The rich physics of whispering gallery modes (WGMs) which are being offered on optical microcavities and optical microresonators have attracted plenty of interests in the research community in the past few decades, particularly in the field of advanced optoelectronics devices [1,2]. The main attractions of dielectric optical microresonators lies in their ability to confine light with high quality Q factor, high finesse but within a small mode volume. Light which are confined on the surface of the microresonators would interact with any bound particles in thousands, millions or billions number of times (depending on the surface roughness) for enhanced light-matter interaction. Among the reported optical WGM microresonators with high finesse and Q-factor values is the microsphere [3], microtoroid [4] and microdisc [5]. While the aforementioned microresonators would confine light only along their azimuthal axis, optical bottle microresonators (BMRs) attain the ability to confine light in a 3-dimensional manner along their azimuthal and axial axis [6]. High number of accessible modes are also easily excited due to its spheroidal profile with Q-factor values of up to 10^8 [7]. Additionally, optical fibre stems which comes naturally with BMRs adds to their easy-handling and practicality such as for strain sensing applications [8].

In contrast to the superior capabilities of dielectric WGM microresonators, the performance of surface plasmon polarition (SPP) cavities have been previously demonstrated to attain low Q-factor values which are mainly due to Ohmic loses in metals [9]. In order to overcome the drawback, a hybrid WGM-plasmonics microresonator was proposed and demonstrated where a high Q-factor dielectric microdisc was partially coated with a noble metal [10]. Since then, other types of hybrid WGM-SPP microcavity resonators has been demonstratred as well including microtoroid [11], microcylinder [12] and BMR [13]. In this paper, we present our study on the performance of a hybrid WGM-SPP BMR with a meniscus thin gold (Au) film. Compared to previous hybrid WGM-SPP cavities which attain constant thickness of noble thin metal-film, we report a hybrid BMR with varying thin metal-film thickness and tapered edges covering half of the BMR surface. *Q*-factor values of composite TE and TM photonics-plasmonics WGM in the range of 1800 and 2700 was calculated, respectively.

2. Experimental Setup

A "soften-and-compress" technique which utilizes an optical fusion splicer was applied on a small section of a continuous commercial single-mode fibre (SMF) in order to generate the spheroidal profile of the BMR [14]. The fabricated BMR attains bottle diameter of ~ 180 μm, neck-to-neck length of ~ 400 μm and fibre stem diameters of 125 μm. For SPPs generation, the BMR was then coated with a thin layer of Au film through metal-evaporation process. The metal deposition process evaporates Au particles from bottom to the top (where the BMR was fixed) in a controlled vacuum chamber system. The combination of this metal deposition process with BMR surface curvature ensures that only half of the BMR surface would be coated with the thin Au-film along with a meniscus profile and tapered edges, as shown in figure 1 (a). The maximum thickness of the Au film on the BMR was measured to be 200 nm. The exposed dielectric part of the hybrid plasmonic bottle microresonator (PBMR) is crucial for light coupling processes.

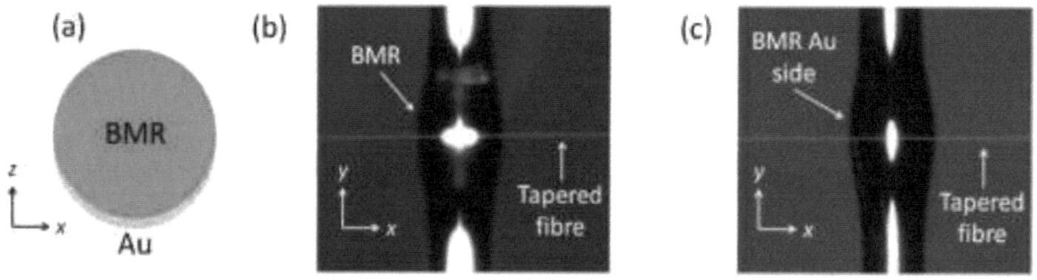

Figure 1. (a) Cross-section schematic diagram of Au metal-film with a meniscus profile on BMR surface, (b) image of the BMR coupled with a tapered optical fibre on the dielectric side and (c) Au coated side

To excite WGMs and SPPs of the hybrid microresonator, tapered optical fibre with 2 μm waist diameter and an adiabatic profile was used to evanescently coupled lasing light into the dielectric part of the microcavity [15]. A polarization-resolved experimental setup which incorporates free-space optics was used to characterize the hybrid PBMR [16]. The PBMR was positioned on precise optical fibre rotating mounts on its two sides by firmly clamping the fibre stems. The rotating mounts was secured on top of two optical table micro-stages for practical handling of the microresonator. The rotating mounts were utilized in order to precisely determine the angle of rotation of the PBMR so that varying tapered fibre coupling arrangements relative to the thickness of the Au would occur. Lasing light with fixed power of – 7 dBm was launched into the SMF between 1520 – 1530 nm wavelength by using a tunable laser source (TLS) (Agilent 81600B). The loss of the free-space optics was determined to be < 8 dB during the entire experimental procedure. The polarization of the input light was tuned by using an optical fibre polarization controller (PC) and the polarization state of the coupled light was determined by a using a high extinction ratio (> 45 dB) polarizer lens. The polarizer lens was placed in a precise rotation mount for straightforward 0° (TM) and 90° (TE) rotation. The other end of the SMF was then spliced to an adiabatic tapered optical fibre with its waist section being placed on top of and in contact with the PBMR. For precise coupling arrangements, the tapered optical fibre was fixed on an optical micro-stage during the entire experimental procedure. The other end of the tapered fibre was cleaved at its full 125 μm diameter for free-space output light collection. An objective lens L1 was brought close to the cleaved optical fibre in order to collimate the output light and to guide it to pass through the polarizer lens for the input light polarization tuning process. The polarized light was then focused down by another objective lens L2 into a multimode optical fibre (MMF). The other end of the MMF was connected to a high sensitivity (-80 dBm to + 10 dBm) optical power meter (Agilent 81635A) for the output light collection. To attain precise TE- and TM-excited modes, a sharp contrast between the resonances was determined during the polarization tuning process.

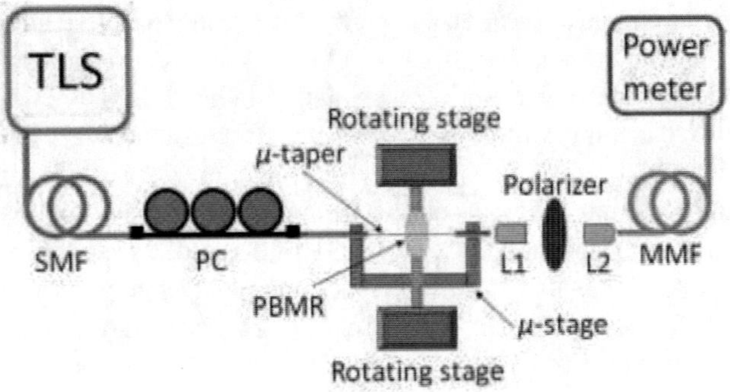

Figure 2. Polarization-resolved experimental setup

3. Results and Discussion

The output transmitted spectra of the PBMR resonances with various coupling arrangement is shown in figure 3. Each dip in the spectrum represents either a single resonance or resonances of a family group-mode. From the output spectrum, the free spectral range of the PBMR was measured to be ~ 3.0 nm. For TE excited WGMs, two specific resonances, G (green dashed-line) and H (blue dashed-line), were monitored during the experiment. The rotation of the BMR θ_{BMR} was first set to be at 0° angle relative to the position of the tapered optical fibre (which is the furthest from the thickest part of the Au film). Resonance G and H were first measured to attain similar coupling strength in the range of 6 dB with such coupling arrangement. With $\theta_{BMR} = 23°$, the coupling strength of resonance G became weaker to ~ 4 dB while the coupling strength of resonance H increased to ~ 8 dB. Interestingly, with $\theta_{BMR} = 113°$, many other distinguishable resonance dips appeared with neither resonance G nor H attaining the highest coupling strength. Placing the tapered fibre on the Au coated part of the PBMR would not couple any light into the microcavity and all of the launched power would pass through the tapered fibre, as could be seen with coupling arrangement $\theta_{BMR} = 180°$. Low coupling strength of resonance G and H were observed with $\theta_{BMR} = 315°$ with many other weak resonance dips throughout the spectrum wavelength.

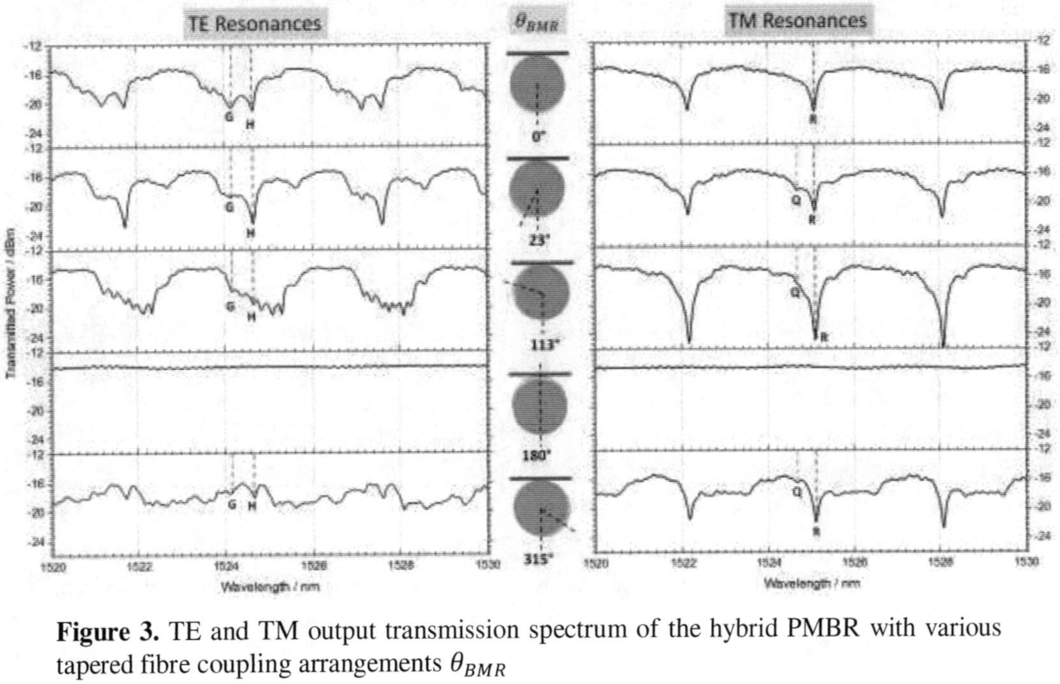

Figure 3. TE and TM output transmission spectrum of the hybrid PMBR with various tapered fibre coupling arrangements θ_{BMR}

It should be stressed that since SPPs are transverse magnetic in nature, TM-polarized light are required to excite plasmonics wave on any noble metal surface. With TM-polarized light launched into the cavity of the PBMR, two main resonances, Q (orange dashed-line) and R (red dashed-line) were monitored with various coupling arrangements. With $\theta_{BMR} = 0°$, only resonance R was observed on the output spectrum with ~ 8 dB coupling strength. Resonance Q only weakly appeared with $\theta_{BMR} = 23°$ while resonance R coupling strength weakens. Strong and distinguishable TM-polarized R resonances were generated with $\theta_{BMR} = 113°$ with coupling strength of > 10 dB along with weak Q resonances. This would be the most promising coupling arrangement to study upon for an enhanced plasmonics wave generation on a BMR. As with the previous observations, no light was coupled into the cavity of the PBMR with $\theta_{BMR} = 180°$ and resonances with moderate coupling strength were generated with $\theta_{BMR} = 315°$. For TM excited resonances, the highest composite Q-factor value calculated $(Q = \lambda/\Delta\lambda)$ was in the range of 2700 at $\theta_{BMR} = 113°$ and for TE excited resonances, the highest composite Q-factor value calculated was in the range of 1800 at $\theta_{BMR} = 23°$.

From the recorded output spectrum, it could be stated that TE- and TM-resonances of the PBMR would vary with different coupling arrangements of the tapered fibre relative to the position of the meniscus thin metal-film. To gain better insight of the obtained result, Lorentzian fittings were performed using Origin computer software on the output spectrum with $\theta_{BMR} = 113°$, where most of the TE resonances appeared and strongest coupling efficiencies of TM resonances were observed. As shown in figure 4, in order to obtain the most accurate fitting, the cumulative fit peak (cyan solid line) of the Lorentzian must be as matching as possible with the real measured spectrum (black solid line). With TE-polarized input light, eight main individual resonances were observed within the family group-mode centring at 1522 nm. However, due to vernier effects, only seven main resonances were observed at family group-mode centring at 1525 and 1528 nm. The resonances of the longer wavelengths in these family group-modes were very closely excited to one another resulting them to be indistinguishable in the fitting. Lorentzian fitting on the TM-polarized output spectrum yields three main resonances with one weak resonance (orange solid line) appearing on the shorter wavelength of the family group-mode. At family group-mode centring 1528 nm, this weak resonance could not be fitted using the software as the vernier effect shown by the measured spectrum (solid black line) has separate it to even weaker resonances with weak resonance dips. The strongest resonance (resonance R) presented by the red solid line attain the highest coupling efficiency of < 9 dB, slightly lower than the maximum reached dip. The reason for this lower coupling strength fitting is because multiple resonance dips comprising of individual resonances (as in the actual family group-mode measured data) would slightly lower the output power at a specific resonance dip which makes an individual resonance appearing to attain higher coupling efficiency.

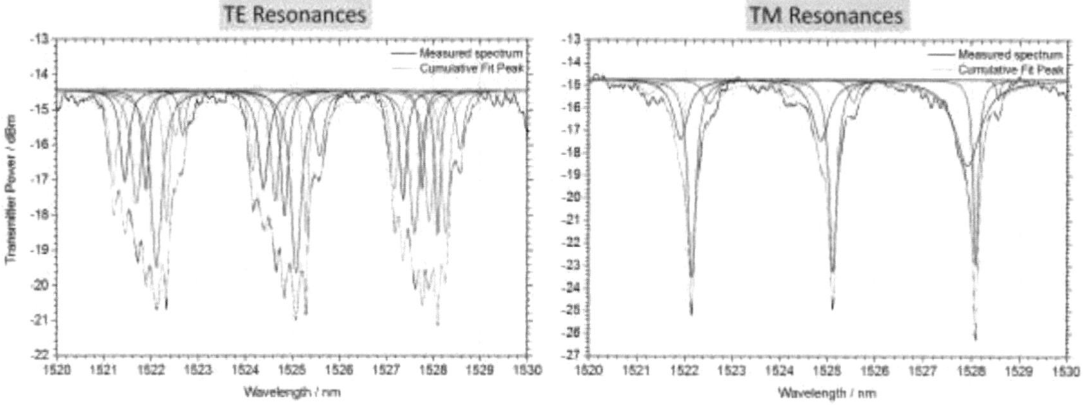

Figure 4. Measured spectrum, Lorentzian fittings and cumulative fit peak of TE and TM mode resonances of the PBMR output transmission spectrum with fixed coupling arrangements $\theta_{BMR} = 113°$

4. Conclusion

In summary, excitation of WGM-SPP resonances utilizing an optical BMR has been successfully demonstrated. The proposed hybrid PBMR attains a thin meniscus Au metal-film of 200 nm maximum thickness along with tapered edges. Different coupling arrangements of the tapered fibre relative to the position of the meniscus Au film could be realized with different θ_{BMR} performed. Coupling strengths of the excited resonances would vary with different coupling arrangements both for TE- and TM-excited resonances. Lorentzian fitting reveals that TE and TM excited resonances actually comprise of eight and three main individual resonances, respectively. Composite resonances Q-factor value was calculated to be in the range of 1800 for TE- and 2700 for TM-mode resonances. The varying excitation strengths of the proposed hybrid PBMR resonances would be beneficial for future applications of optoelectronics plasmonics devices.

References

[1] Vahala K J 2003 Optical microcavities *Nature* **424** 839
[2] Foreman M R, Swaim J D and Vollmer F 2015 Whispering gallery mode sensors *Adv. in Opt. Photon.* **7** 168
[3] Gorodetsky M L, Savchenkov A A and Ilchenko V S 2000 Ultimate Q of optical microsphere resonators *Opt. Lett.* **21** 453
[4] Armani D K, Kippenberg T J, Spillane S M and Vahala K J 2003 Ultra-high-Q toroid microcavity on a chip *Nature* **421** 925
[5] Little B E, Foresi J S, Steinmeyer G, Thoen E R, Chu S T, Haus H A, Ippen E P, Kimerling L C and Greene W 1998 Ultra-compact Si-SiO2 microring resonator optical channel dropping filters *IEEE Photon. Technol. Lett.* **10** 549
[6] Sumetsky M 2004 Whispering-gallery-bottle microcavities: the three dimensional etalon *Opt. Lett.* **29** 8
[7] Pöllinger M, O'Shea D, Warken F and Rauschenbeutel A 2009 Ultra-high-Q tunable whispering-gallery-mode microresonator *Phys. Rev. Lett.* **103** 053901
[8] Mohd Nasir M N, Murugan G S and Zervas M N 2019 Tunable 'shallow' microbottle resonators *IEEE Photon. Technol. Lett.* **31** 849
[9] Boriskina S V, Cooper T A, Zeng L, Ni G, Tong J K, Tsurimaki Y, Huang Y, Meroueh L, Mahan G and Chen G 2017 Losses in plasmonics: from mitigating energy dissipation to embracing loss-enabled functionalities *Adv. Opt. Photon.* **9** 775
[10] Min B, Ostby E, Sorger V, Ulin-Avila E, Yang L, Zhang X, and Vahala K J 2008 High-Q surface-plasmon-polariton whispering-gallery microcavity *Nature* **457** 455
[11] Xiao Y F, Zou C L, Li B B, Li Y, Dong C H, Han Z F and Gong Q 2010 High-Q exterior whispering gallery modes in a metal-coated microresonator *Phys. Rev. Lett.* **105** 153902
[12] Zhang X, Ma Z, Yu H, Guo X, Ma Y, and Tong L 2011 Plasmonic resonance of whispering gallery modes in an Au cylinder *Opt. Exp.* **19** 3902
[13] Rottler A, Harland M, Bröll M, Klingbeil M, Ehlermann J and Mendach S 2013 High-*Q* hybrid plasmon-photon modes in a bottle resonator realized with a silver-coated glass fiber with a varying diameter *Phys. Rev. Lett.* **111** 253901
[14] Murugan G S, Wilkinson J S and Zervas M N 2009 Selective excitation of whispering gallery modes in a novel bottle microresonator *Opt. Exp.* **17** 11916
[15] Mohd Nasir M N, Murugan G S and Zervas M N 2016 Spectral cleaning and output modal transformations in whispering-gallery modes resonators *J. Opt. Soc. Am. B* **33** 1963
[16] Mohd Nasir M N, Gorajoobi S B, Murugan G S and Zervas M N 2019 Polarization effects in optical microresonators *J. Opt. Soc. Am. B* **36** 705

Q-switched Ytterbium-doped fibre laser using an 8 cm long Hafnium bismuth erbium co-doped fibre saturable absorber

M F A Rahman[1*], P H Reddy[5], A Ahmad[1], A A Latiff[2], M F Baharom[3], and S W Harun[4]

[1] Fakulti Teknologi Kejuruteraan Elektrikal dan Elektronik, Universiti Teknikal Malaysia Melaka, Hang Tuah Jaya, 76100 Durian Tunggal, Melaka, Malaysia
[2] Fakulti Kejuruteraan Elektronik dan Kejuruteraan Komputer, Universiti Teknikal Malaysia Melaka, Hang Tuah Jaya, 76100 Durian Tunggal, Melaka, Malaysia
[3] Fakulti Kejuruteraan Elektrik, Universiti Teknikal Malaysia Melaka, Hang Tuah Jaya, 76100 Durian Tunggal, Melaka, Malaysia
[4] Department of Electrical Engineering, University of Malaya, 50603, Kuala Lumpur, Malaysia
[5] Fiber Optics and Photonics Division, CSIR-Central Glass and Ceramic Research Institute, 196, Raja S.C. Mullick Road, Kolkata 700032, India

[*]Corresponding author's email: mfauziar@utem.edu.my

Abstract. In this paper, we present a Q-switched fibre laser at 1069 nm which is induced by an 8 cm long Hafnium bismuth erbium co-doped fibre (HBEDF) saturable absorber (SA). The pulsating laser has a maximum repetition rate of 67 kHz at 175 mW pump power. We obtained the narrowest pulse width of 3.48 μs, the maximum pulse energy of 70.2 nJ, the maximum output power of 4.7 mW and the maximum peak power of 20.1 mW. The Q-switched laser is simple and may found practical applications in medicine and remote sensing.

Keywords: saturable absorber, pulsed fibre laser, passive method, ytterbium-doped fibre laser

1. Introduction

Q-switched fibre laser via passive technique has been quite popular in recent years due to its small size and simple fabrication, making it suitable for biomedical, material processing, and remote sensing applications [1, 2]. Semiconductor saturable absorber mirror (SESAM)[3], carbon nanotubes (CNTs) [4], and graphene [5, 6] are some of the early SAs with excellent performance. Later, many other SAs, have been demonstrated [7, 8]. However, most of the SAs are film-based since they are easy to prepare [9-14]. However, the film SAs have some issues, as they are easy to get oxidized, degraded, and burnt after certain days of usage.

Recently, our lab has demonstrated a few Q-switched lasers based on fibre SAs [15-19]. The fibre SA is flexible, simple, and has good similarities to the active medium, making it durable and easy to maintain.

This paper demonstrates a stable Q-switched ytterbium-doped fibre laser (YDFL) that uses an 8 cm long HBDEF as the passive SA. The Q-switched laser can be tuned from 41 kHz to 67 kHz by adjusting the pump power within the range of 106 mW to 175 mW.

2. Methodology

2.1. Saturable absorber preparation

The HBEDF was made by incorporating Hafnium (Hf), Bismuth (Bi) and Erbium (Er) element in a silica glass via a modified chemical vapour deposition process. The doping level of the fabricated fibre was checked through an electron-probe microanalysis. The result indicates the corresponding elements weight percentage as follows; 1.23 wt % of Er, 2.2 wt % of Hf and 0.035 wt % of Bi.

The fabricated fibre was cut into a short segment of 8 cm. Both ends of the fibre were then spliced with a single-mode fibre (SMF28), which gives a final SA dimension of (8 cm SMF28 – 8 cm HBEDF – 8 cm SMF28). However, for simplicity, we regard the fibre SA as 8cm HBDEF SA. The linear absorption profile (LAP) was measured by placing the fibre SA between a white light source (WLS) and an optical spectrum analyzer (OSA) by considering the absorption value with and without the fibre SA. We noticed sufficient absorption of 3dB at the Q-switched operating wavelength (1069 nm YDFL), as depicted in **Figure 1**.

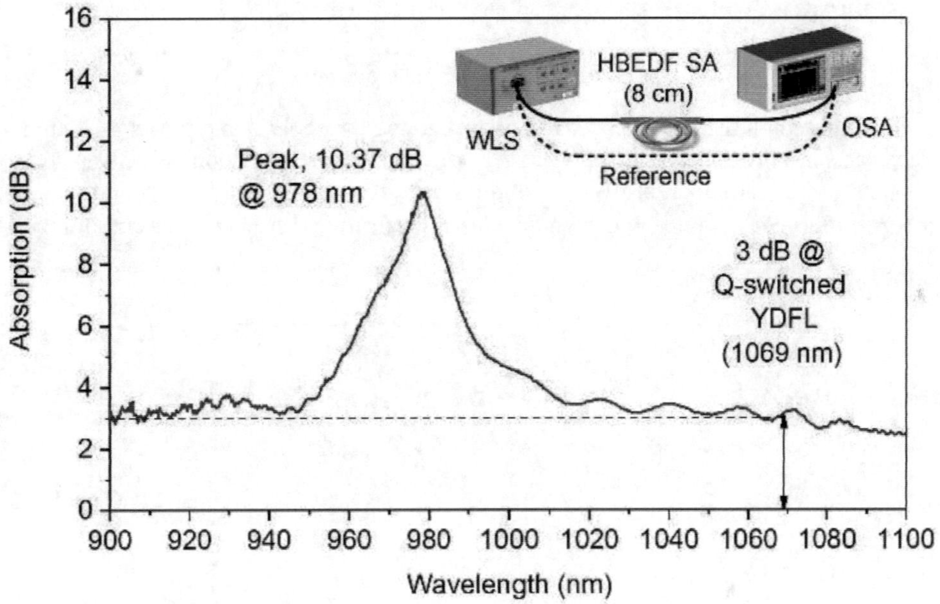

Figure 1. The fibre SA linear absorption profile

2.2. Laser cavity configuration

The laser cavity consists of a laser diode pump (980 nm), a single-core ytterbium-doped fibre (YDF, 1.5 m), a 980/1060 wavelength division multiplexer (WDM), a 3 dB optical coupler, and the HBDEF SA. The arrangement of the laser cavity is depicted in **Figure 2**. The isolator keeps the unidirectional light propagation. The 3 dB coupler splits the light into a 50:50 ratio. Half of it is channelled into optical measurement devices such as OSA, a digital oscilloscope and a radio frequency spectrum analyzer (RFSA).

Journal of Physics: Conference Series **2075** (2021) 012020 doi:10.1088/1742-6596/2075/1/012020

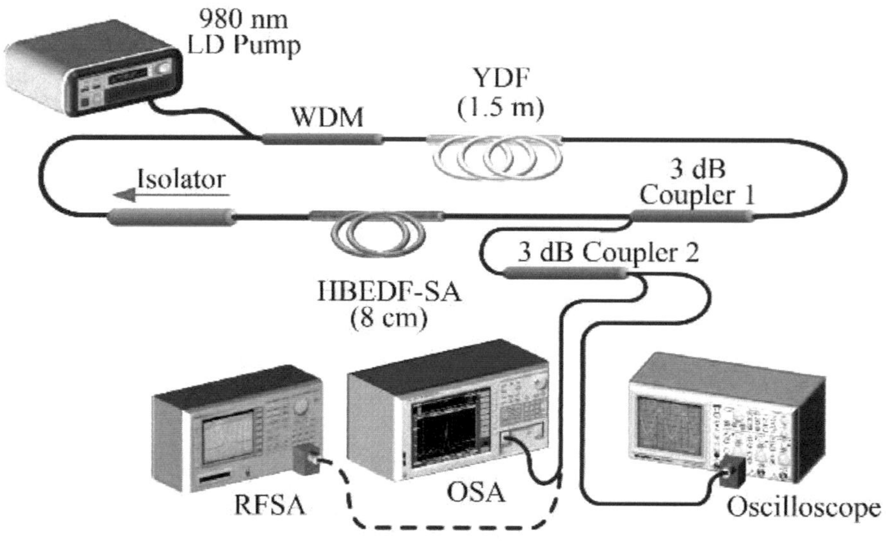

Figure 2. HBDEF SA based Q-switched laser cavity configuration

3. Results and discussion

The Q-switched YDFL based on the HDEF SA starts to appear at a threshold pump power of 106 mW. The pulsed laser has an initial repetition rate of 41 kHz. The pulse frequency monotonically increases to 67 kHz as the pump power rises to 175 mW. Beyond the 175 mW, the pulsed laser collapses into a continuous laser (CW). **Figure 3** shows the Q-switched laser spectrum, centred (λ_c) at 1069 nm with a peak intensity of -15.2 dBm.

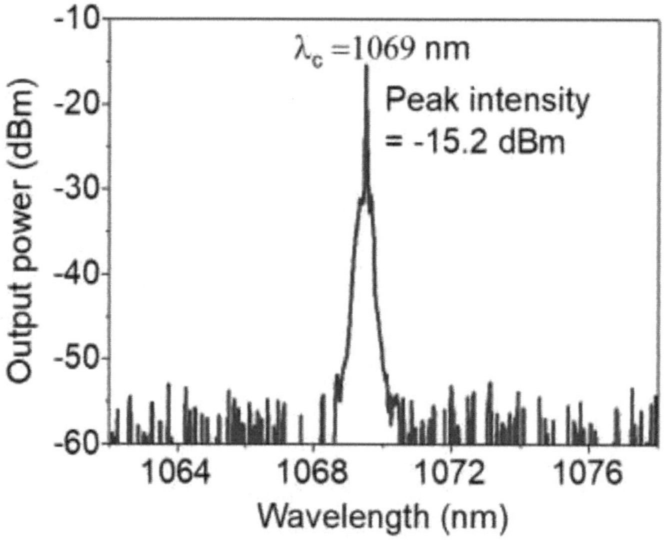

Figure 3. HBDEF SA based Q-switched laser spectrum at the 175 mW pump power

The Q-switched laser viewed in the time domain is shown in **Figure 4**. As illustrated, the generated pulse train has a frequency of 67 kHz, which reflects the pulse period of 14.9 μs. The pulse duration (τ) measured at the full width at half maximum (FWHM) is observed to be around 3.48 μs.

We also measured the radio frequency spectrum of the Q-switched laser. We observed consistent frequency harmonics, with a fundamental frequency (FF) of 67 kHz, the same as the one revealed in the time domain. Furthermore, the signal to noise ratio (SNR=48 dB) is considerably high, suggesting that the generated Q-switched laser is stable.

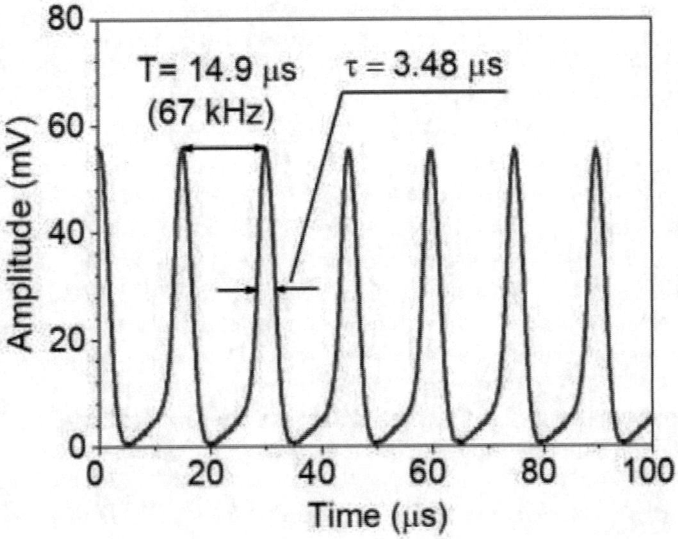

Figure 4. HBDEF SA based Q-switched pulse train at the 175 pump power

4. Conclusion

We demonstrated a Q-switched fibre laser at 1069 nm by integrating an 8 cm long HBDEF SA. The pulsed laser has the shortest pulse width of 3.48 μs and the maximum pulse frequency of 67 kHz at the 175-mW pump power. The maximum pulse energy is 70.2 nJ. The Q-switched laser is simple, flexible and can be used in many applications, including biomedical and material processing.

References

[1] Sun D, Xu X, Chen J, Shi W, Sun L, Chu S and Ruan S 2021 Passively Q-switched ytterbium-doped fiber laser based on a SWCNT@ AFI saturable absorber *Optics & Laser Technology* **136** 106781

[2] Al-Hiti A S, Al-Masoodi A H, Wong W R and Harun S W 2020 Generation of passively Q-switched ytterbium laser by using tungsten tri-oxide film absorber *IET Optoelectronics* **14** 278-84

[3] Keller U, Miller D, Boyd G, Chiu T, Ferguson J and Asom M 1992 Solid-state low-loss intracavity saturable absorber for Nd:YLF lasers: an antiresonant semiconductor Fabry–Perot saturable absorber *Optics letters* **17** 505-7

[4] Song Y W, Yamashita S, Goh C S and Set S Y 2007 Carbon nanotube mode lockers with enhanced nonlinearity via evanescent field interaction in D-shaped fibers *Optics Letters* **32** 148-50

[5] Sun Z P, Hasan T, Torrisi F, Popa D, Privitera G, Wang F Q, Bonaccorso F, Basko D M and Ferrari A C 2010 Graphene Mode-Locked Ultrafast Laser *Acs Nano* **4** 803-10

[6] Sobon G and Sotor J 2016 Recent Advances in Ultrafast Fiber Lasers Mode-locked with Graphenebased Saturable Absorbers *Current Nanoscience* **12** 291-8

[7] Luo Z, Huang Y, Weng J, Cheng H, Lin Z, Xu B, Cai Z and Xu H 2013 1.06 μm Q-switched ytterbium-doped fiber laser using few-layer topological insulator Bi 2 Se 3 as a saturable absorber *Optics express* **21** 29516-22
[8] Tey K, Rahman M, Rosol A, Rusdi M, Mahyuddin M, Harun S and Latiff A 2017 Q-switched double-clad Ytterbium-doped fiber laser using MoS2 flakes saturable absorber. In: *IOP Conference Series: Materials Science and Engineering*: IOP Publishing) p 012054
[9] Rahman M, Mahyuddin M, Latiff A, Paul M, Dhar A, Das S, Yupapin P, Yasin M and Harun S 2019 Holmium based nanoseconds pulsed fibre laser generation in the 2-micron region *Optik* **195** 163157
[10] Rahman M, Zhalilah M, Latiff A, Rosol A, Lokman M, Bushroa A R, Dimyati K and Harun S W 2018 Pure antimony film as saturable absorber for Q-switched erbium-doped fiber laser *Journal of Modern Optics* **65** 811-7
[11] Rusdi M, Rosol A, Rahman M, Mahyuddin M, Latiff A, Ahmad H, Harun S and Yasin M 2019 Q-switched and mode-locked thulium doped fiber lasers with nickel oxide film saturable absorber *Optics Communications* **447** 6-12
[12] Rahman M F A, Latiff A A, Zaidi U Z M, Rusdi M F M, Rosol A H A, Bushroa A R, Dimyati K and Harun S W 2018 Q-switched and mode-locked thulium-doped fiber laser with pure Antimony film Saturable absorber *Optics Communications* **421** 99-104
[13] Ikhwan M, Rahman M, Fatheli M, Sammsamnun F, Haroon H and Latiff A 2021 Q-switched fibre laser using Samarium oxide for C-band region *Journal of Optoelectronics and Advanced Materials* **23** 307-12
[14] Rahman M, Rusdi M, Harun S and Yasin M 2018 Q-Switched Thulium-Doped Fiber Laser Using Vanadium Oxide Saturable Absorber *Nonlinear Optics, Quantum Optics: Concepts in Modern Optics* **48**
[15] Rahman M, Dhar A, Das S, Dutta D, Paul M, Rusdi M, Latiff A, Dimyati K and Harun S 2018 An 8 cm long holmium-doped fiber saturable absorber for Q-switched fiber laser generation at 2-μm region *Optical Fiber Technology* **43** 67-71
[16] Rahman M, Latiff A, Reddy P, Das S, Dhar A, Paul M and Harun S 2019 Passively Q-switched fiber laser utilizing new hafnium–bismuth–erbium co-doped fiber as saturable absorber *Indian Journal of Physics* 1-5
[17] Rahman M, Reddy P, Paul M C, Das S, Dhar A, Baharom M, Latiff A, Rusdi M, Wang P and Dimyati K 2019 Titanium dioxide fiber saturable absorber for Q-switched fiber laser generation in the 1-micrometer region *Applied optics* **58** 3495-500
[18] Rahman M F A, Dhar A, Das S, Dutta D, Paul M C, Rusdi M F M, Latiff A A, Dimyati K and Harun S W 2018 An 8 cm long holmium-doped fiber saturable absorber for Q-switched fiber laser generation at 2-μm region *Optical Fiber Technology* **43** 67-71
[19] Rahman M F A, Latiff A A, Rusdi M F M, Dimyati K and Harun S W 2019 Passively Q-switched fibre laser utilizing erbium-doped fibre saturable absorber for operation in C-band region *Journal of Modern Optics* **66** 235-9

Acknowledgement
The research was supported by the Ministry of Education (MOE) through Fundamental Research Grant Scheme (FRGS/1/2021/FTKEE/F00489). In addition, the authors are grateful to Universiti Teknikal Malaysia Melaka for the research and technical support.

Optical Microfiber Sensor : A Review

Mohd Hafiz Jali[1]*, Hazli Rafis Abdul Rahim[2], Md Ashadi Md Johari[3], Mohamad Faizal Baharom[1], Aminah Ahmad[3], Haziezol Helmi Mohd Yusof[2] and Sulaiman Wadi Harun[4]

[1]Faculty of Electrical Engineering, Universiti Teknikal Malaysia Melaka, 76100 Melaka, Malaysia
[2]Faculty of Electronic and Computer Engineering, Universiti Teknikal Malaysia Melaka, 76100 Melaka, Malaysia
[3]Faculty of Electrical and Electronic Engineering Technology, Universiti Teknikal Malaysia Melaka, 76100 Melaka, Malaysia
[4]Department of Electrical Engineering, University of Malaya, 50603 Kuala Lumpur, Malaysia

*Corresponding author's email: mohd.hafiz@utem.edu.my

Abstract. Due to numerous benefits such as geometrical simplicity, compact size, high sensitivity, broad detection range, low noise, and high accuracy, optical devices have attracted a lot of interest for sensing applications. It is critical in a variety of sectors, including cultural relic preservation, warehouse products maintenance, manufacturing process control, semiconductor, agriculture, food production storage, environmental control, health industries, chemical and home improvement. It outperforms its electronic equivalent owing to its capacity to function in tough and demanding situations such as combustible surroundings, greater pressure and temperature levels, and the ability to send signals over long distances without electromagnetic interference. Optical fiber sensors are classified based on their operating principles such as interferometers, fiber Bragg gratings (FBG), resonators and whispering galleries modes (WGM). This paper presents a comprehensive review related to the optical microfiber sensor such as its properties, fabrication techniques, evanescent wave, optical micro resonators and recent study on the application of microfiber towards humidity sensing. This review could be beneficial to help other researchers to gain greater view in the field of optical microfiber sensor..

Keywords: microfiber, optical sensor, evanescent wave

1. Introduction

During the last century, optical fibers were widely used in the field of telecommunications due to their special attributes such as low transmission loss, high bandwidth and multiplexing capability. Recently the optical fiber application has also shifted to other areas such as sensing to take advantage of the special features of it. Even though the price of the established electronic sensors counterpart is more competitive, optical sensors could find the target application by utilizing their unique features. Some of the unique features are low losses, remote sensing, immune to electromagnetic interference, no electrical biasing is required to guide light and the viability to use in an environment which has explosion risk. Up to date, optical fiber sensor has gained tremendous interests for various bio-chemical sensing applications such as ethanol, sodium hypochlorite, formaldehyde and human chorionic gonadotropin [1-6]. It is driven by the rapid development of micro/nanotechnology and the need to reduce the size of the sensor while maintaining the key requirement of the sensor such as low power consumption, faster response, better spatial resolution and higher sensitivity [7, 8].

Microfibers have unique properties like large optical confinement, configurability, flexibility, strong optical confinement, and large evanescent field, making them ideal for physical sensing applications like ultra-sensitive surface absorption spectroscopy, hydrogen detection, chemical and refractive index sensors [9, 10]. Due to large evanescent wave propagates outward from the microfiber, it is extremely sensitive to changes in the ambient refractive index. When the proportion of power transmitted in the evanescent field increases, the refractive index of the surrounding material rises. It exhibits good evanescent coupling with other waveguides such as metal, semiconductor and substrate. Thus, high fractional evanescent field allow great responses towards humidity sensing [9].

The type and number of modes propagating through the fiber could be determined by the physical features of the fiber such as RI, core diameter and operating wavelength. Most portion of light energy resides inside the fiber while some portion infiltrates the clad. The small portion decays exponentially into the core-cladding edge. The low amplitude of the evanescent field in a normal single mode fiber (SMF) could be enhanced by tapering in order to increase interaction of analyte around the taper region with the transmitted light [11].

2. Properties

Microfiber offers several interesting optical properties such as strong evanescent field. Fractional power spreads in the evanescent field at the outer physical boundary of the small radii microfiber. This property is essential to fabricate high quality factor (Q) resonators and to launch light into high Q micro-resonators [12]. Microfiber also exhibits strong near-field interaction with its surrounding [13]. It has good evanescent coupling between the microfiber and the other waveguides such as substrates, semiconductor and planar waveguide. This lead to plenty of optical devices such as resonators, sensors and lasers [14]

Propagation loss also is an important property for microfiber. The propagation loss is due to several factors such as cracks, impurities connected to the micro/nano fiber surface and surface imperfections [12]. It increases as the microfiber radii decreases [15]. Theoretical studies on non-adiabatic intermodal transitions were established to examine the minimum microfiber waist diameter that can propagate signal. The transmission mode would disappear at a threshold rate approximately at a value lesser than the radiation wavelength [16]. It is also found that the small size microfiber degrades faster in air when there is a generation of cracks at the facet due to the water absorption [17].

Besides that, microfiber is very sensitive to momentum change of photon guide through by mechanical displacement or vibration mass due to its small mass [13] This enable a development of compact opto-mechanical component/devices [18]. It also allows a low loss when light passing through sharp bends. Therefore microfiber could realize a compact devices with lower power consumption, faster response and smaller footprints [13, 19]. Microfiber is made-up by extending the optical fibers until reaching the desired waist diameter while keeping the original fiber sizes at their input and output ends. This eases a low loss splicing to interconnect with other standard size optical fibers. It also could be bent into a small bending radius to produce a compact devices [12]. The small waist fiber also has a large fraction of power transmitted outside the microfiber and overlaps with the external elements. Any changes in surrounding properties result in a change in output as shown in Figure 1 [20].

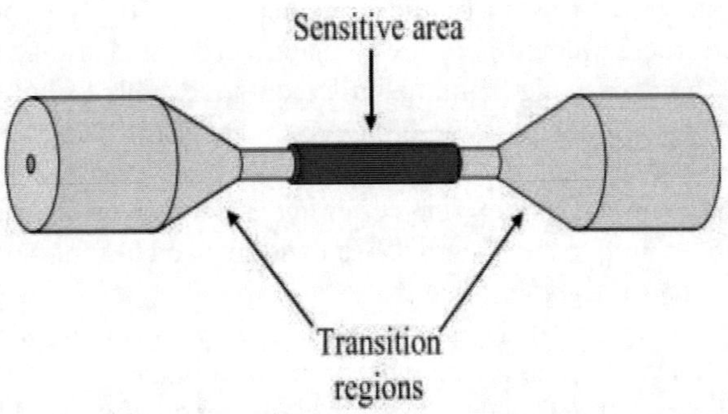

Figure 1. Sensitive area has a large fraction of power propagating to interact with the surrounding environment [20]

3. Fabrication techniques

Tapering of a SMF is a process of decreasing the diameter of the cladding (together with the core) by pulling the fiber's end while heating the waist of the fiber. During the tapering process, the light propagates inside the core infiltrates to the cladding. This cause them to acts as a new core and the external element as a new cladding [21]. In order to achieve high signal to noise ratio and low optical loss criteria, surface smoothness and geometric uniformity during fabrication procedure of the microfiber is crucially essential [13].

The tapered fiber adheres to adiabatic criteria if the main portion of power uncouples with higher order modes and sustain in the fundamental mode as the power propagate along the taper. It could be realized by demonstrating a tiny taper angle to the relative local change of the taper radius [21]. When the fiber was pulled during the tapering process, the radius of the fiber decreases and the light extrude from the core to the cladding and propagates throughout the taper region. Thus, the core, cladding and the air would influence the mode. The mode evolution inside the tapered fiber or microfiber is highly related to the shape of the taper. If the taper is too steep, a non-adiabaticity will occur that cause low transmission. The mode propagation become more adiabatic when the tapering angle reduced [22]. Figure 2 (i) shows the different shape between adiabatic and non-adiabatic tapered fiber.

There are several tapering methods to fabricate glass microfiber. The first technique is self-modulated taper drawing. It requires two-step process: first the waist diameter of the single mode fiber is reduced to a micro-meters using standard flame brushing technique. Then the microfiber is broken into two parts and one of the fiber pigtail is enfolded onto a small hot sapphire rod and the microfiber is further drawn into sub-micron diameter. Subsequently, the tip of the sapphire is heated using a flame at a distance from the fiber to stabilize the distribution of the temperature and trap the heat in a small volume [12]. Even though this technique has a quite complex fabrication process, it is capable to produce a tapered fiber with radius as small as 10 nm [23]. Second method is flame brushing technique as shown in Figure 2 (ii). This technique was initially invent to manufacture couplers [24]. It is conducted by applying small flame movement under a stretched optical fiber. The optical fiber extremities and burner are hold by stages and connected to the computer. An highly accurate microfiber could be fabricated by precisely controlled the flame movement [25]. It could produce micro/nanofibers with radius as small as 30 nm and longest nanowires with less measured loss [17]. Besides that this technique allow the microfiber with both ends pigtailed which is essential in practical applications that have issue on connectivity [12].

The flame brushing technique has been upgraded into modified flame brushing technique. This technique has set-up resembles similar as conventional flame brushing technique but the difference is the flame is changed with different heat source which are a micro-heater [26] or a sapphire capillary tube heated by a CO_2 laser beam [27]. Micro-heater is a resistive element which allows temperature to be adjusted by altering the current level. Whereas, temperature control using a sapphire/CO_2 can be change by tuning the degree focusing of the laser beam onto the sapphire tube. This technique has superiority in terms of processing temperature to manufacture wide range of low softening microfibers [12]. Other method is direct drawing from the bulk technique. This technique required a small sapphire rod to be heated and placed in contact with bulk glass. This results in a localized softening to the bulk glass. As the sapphire rod is abruptly removed, micrometric glass filament is formed [12]. It has advantages in term of flexibility and low-cost equipment. However, the uniformity and the diameter of the microfiber are highly tough to retain. It has been widely used to fabricate microfiber from phosphate, tellurite and polymers [28].

Besides of the tapering method, heat sources also play essential role to produce a good quality microfiber. There are four heat sources have been commonly used during the tapering process such as flame [29], fusion splicer [30], CO_2 laser beam [31] and micro furnace [30]. There are two canonical segments in which the first one occurred relatively at long taper waist segment where the fiber is gradually decreases into a small and uniform diameter. While the second conical segment is the waist merging into the original SMF size [21]. The tapered fiber could be formed into different properties and shapes by controlling pulling settings such as pulling speed, length of the heated zone and pulling temperature. The most common heated source is flame-heated as shown in Fig. 6. The fiber is elongated and stretched slowly under a specific pulling force. A hydrogen flame is applied at the centre of the fiber. The process continues until the targeted diameter or length is achieved [13]. The waveguides of the microfiber could be monitored during the tapering process in term of propagation loss, group velocity delay and multimode interference [32].

Conventional flame-heated technique has disadvantages of random turbulence of the oxygen and flame during the burning process. Thus, CO_2 laser has been introduced to solve this issue [13]. Direct laser heating provide a self-regulating control which the stretching process would automatically stops when the diameter of the fiber reaching a desired value [33, 34]. Other technique is electrically heated taper drawing. This technique enables to shape the fiber into various geometries by precisely control the temperature distribution. It also offer an effective technique to draw microfiber with more flexibilities [13, 35].

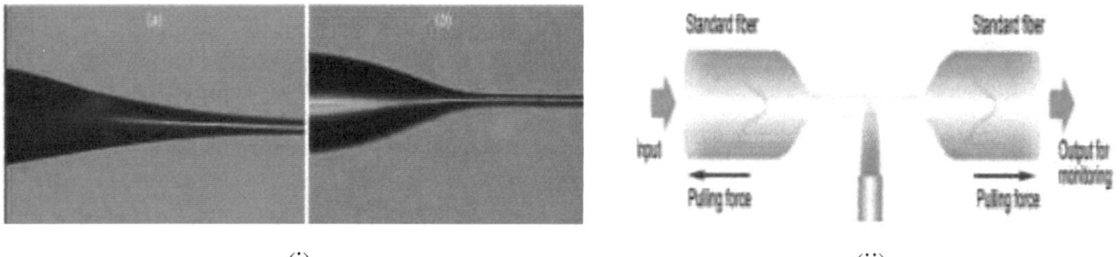

(i) (ii)

Figure 2. (i) Microscopic image of; (a) Adiabatic tapered fiber and (b) Non-adiabatic tapered fiber [21], (ii) Fabrication technique of the microfiber using flame heated source [13]

4. Optical sensor using microfiber

Microfiber offers unique properties, such as visible surface field amplification, strong evanescent field, huge anomalous waveguide dispersion and configurability, intriguing possibilities for sensing applications. Microfiber also exhibits high evanescent coupling with other waveguides including metal, semiconductors, and substrates, as well as significant near-field interaction with its surroundings. Large evanescent field is an essential criteria for sensing application where significant power fraction is needed to interact with surrounding refractive index medium [13].

4.1. Evanescent Wave

A standard silica fiber comprises of two main parts which is core and cladding. The core commonly doped with higher refractive index (RI) to guarantee total internal reflection (TIR) under certain conditions according to Snell law. However, there is always small portion of energy transmission coupled into the cladding modes which is known as evanescent field. [36]. Evanescent wave is leakage/loss of electromagnetic field which represent by small portion of energy that penetrates the edge of core and cladding medium during the TIR. TIR occurred when incident angle (θ_i) is greater than critical angle (θ_c) which cause the light reflection back from core to cladding surface.

Studies show that the fraction of power in the evanescent field increases when normalise wavelength (λ/r) and surrounding refractive index increase [12]. Based on Figure 3, the fraction of power (η_{EF}) depend on the ratio of λ/r. η_{EF} increase monotonically when λ/r increase [37]. Hence, microfiber with smaller radius would produce larger evanescent wave fractional power. A fraction of power propagates in the evanescent field would exist when the diameter of the fiber is comparable to the wavelength. The evanescent field interact with any surrounding environmental change which makes it suitable for sensing application.

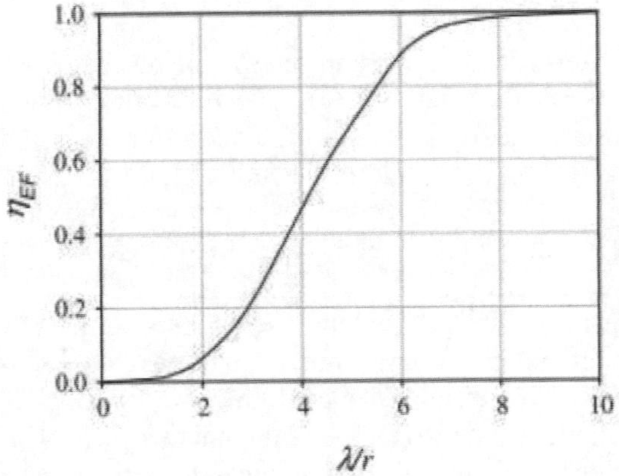

Figure 3. Relationship between fraction of power (η_{EF}) of the silica microfiber and the normalised wavelength (λ/r) ([37]

4.2. Optical microfiber resonators (OMRs)

Microfiber can be employed to manufacture various resonant structures such as micro-loop micro-knot and micro-coil as well as to excite resonant modes such as microspheres, micro-disks, microcapillaries and microbottle. This would produce modes propagating between two adjacents large evanescent field section that overlap and couple to produce a compact resonator [37, 38]. The most important parameter in resonator is quality factor Q. It can be calculated by dividing the wavelength λ with the bandwidth (FWHM) of a resonance in the transmission spectrum as given in (1).

$$Q = \frac{\lambda}{FWHM} \qquad (1)$$

Vienna et al. has used loop and knot resonator as refractometric sensors. The sensors operate by exploiting considerable fraction of the microfiber mode that transmitting in the fluidic channel [39]. Any variation in the analyte refractive index would results in a resonant wavelength shift as shown in Figure 4.

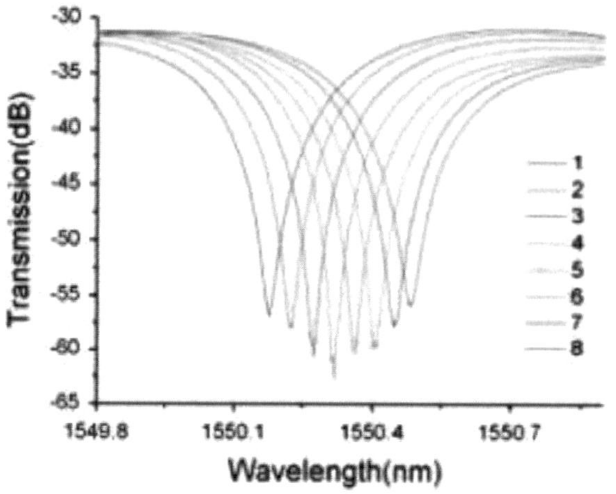

Figure 4. Resonant wavelength shift of loop and knot resonator as refractometric sensors [40]

Xu et al. developed a refractometric sensor using micro-coil resonator with an intrinsic channel for microfluidic applications as shown in Figure 5 (a) [41]. Its working principle is similar as the loop resonator refractometric sensor. It associates with the overlap between the evanescent wave and the analyte during the mode propagating in the microfiber. Resonant wavelength shift would occur if there is any refractive index change [42]. The microcoil resonator was embedded in Teflon and the sensor was exposed to isopropanol and methanol. The resonant wavelength shifts to longer wavelength as the analyte refractive index increases.

The most dominant resonant sensors is based on heterogeneous sensors which exploit microfiber to extract light from high-Q resonators such as microcapillaries, microtoroids, microspheres and microbottle resonator [43]. Coupling efficiencies could achieve up to 90% by properly matching the propagation constant of the mode in the resonator and the microfiber [44]. These high-Q resonators are suitable for evanescent sensing for biological and chemical detection due to its large resonator surface [42, 45]. For instance, microspheres with Q-factor of 2 x 10^6 has been used to detect protein streptavidin [46]. While Noto et al. [47] uses multiple spheres coated with dextran-biotin hydrogel for DNA strands detection. Another work by [48] implement microsphere with radius of 39 um at wavelength of 763 nm to identify influenza A virus as shown Figure 5 (b).

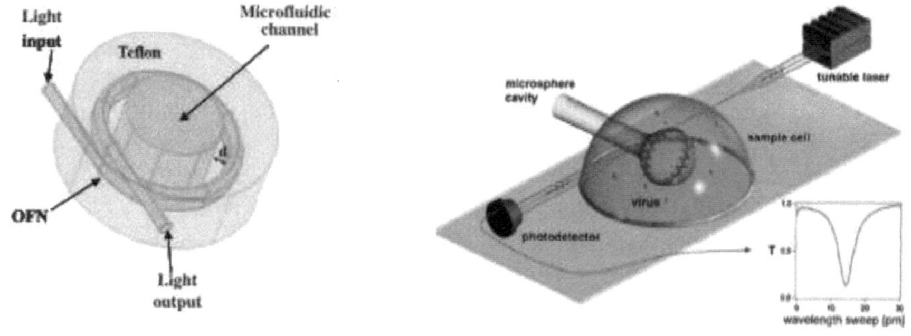

(a) (b)

Figure 5. (a) Microcoil resonator for refractometric sensors using an intrinsic channel for microfluidic applications [41], (b) Heterogeneous sensors which exploit microsphere to detect influenza A virus [48]

5. Recent works on optical microfiber humidity sensing

Humidity is defined as the content of water vapor in the gases. It has been commonly expressed in terms of relative humidity (%RH). It is the ratio of actual vapor pressure at a particular temperature (P_w) divide by the saturation vapor pressure at the same temperature (P_{ws}) as described on formula $\%RH = P_w/P_{ws} \times 100\%$ [49]. Agriculture, meteorology, sterilisers, incubators, textile manufacturing, air conditioning, and chemical gas purification are just a few of the applications. Despite the fact that optical fiber humidity sensors are more expensive than typical electronics sensors, they have advantages such as compact size, the ability to multiplex data from numerous sensors, and insensitivity to electronic magnetic fields [30]. Due to its poor sensitivity and long reaction time, existing percent %RH sensors such as wet and dry bulb psychrometers were not feasible, whilst electronic-based percent %RH sensors suffer from electromagnetic interference [50].

In wet atmosphere, H_2O molecules will compete with the oxygen molecules for the surface reaction sites. As the %RH level increase, the number of oxygen species would continually decrease causing reduction of baseline resistance which translate into lower sensor response [51]. Therefore, the charge carrier transport due to chemisorbed mechanism occur at lower %RH while physisorption based on Grotthuss mechanism takes place at a higher %RH level as depicted in Figure 6 [52, 53].

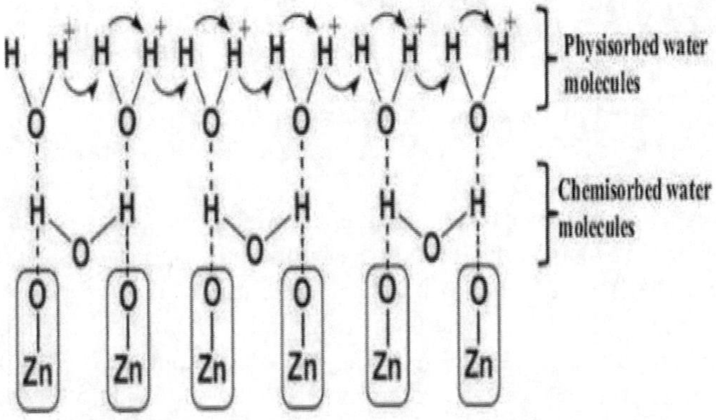

Figure 6. Absorption phenomenon on ZnO [53]

Microfiber bottle resonator, long-period grating coated with gelatin, electrostatic self-assembly of tapered optical fibre, polymethyl methacrylate (PMMA) microfiber doped with agarose gel, silica fiber interferometer, and side polish fibre coated with tungsten disulphide are some of the optical humidity sensing mechanisms that have been proposed [30, 54]. Nanostructure material has attracted attention for use in humidity sensing as compared to the traditional pores ceramic structure because it has hollow structures, ultra-high surface-to-volume ratio, large grain boundary areas and small grain size [55].

The optical humidity sensor working principle could be categorized into several major groups which are based on optical absorption, fiber Bragg gratings (FBG), interferometers and micro-resonators [56]. For optical absorption method, it is based on the reaction between the sensitive material and the evanescent field. This leads to change of transmitted output power. It has advantages on low fabrication cost [57], high reliability and simple setup [58]. However, the drawback of this method is the changes of the transmitted output power could be effected by the undesired factors such as fluctuations of the light source. Several sensitive material have been studied such as zinc oxide [59], tungsten disulfide [60] and reduced graphe oxide (rGO) [61]. For instance, a D-shape optical fiber coated with tungsten disulfide (WS_2) has shown a quite good sensitivity with 0.1213dB/%RH and resolution of 0.475%RH as shown in Figure 7 (a).

Second method is based on FBG which comprises of a periodic perturbation of the refractive index. FBG operated when small amount of light transmitted inside the fiber is reflected at each periodic refraction change [58]. All the reflected signal combine coherently into a al large reflection at specific wavelength when the grating period reaching half of the input's wavelength [62]. The grating phenomenon is described in (2).

$$\lambda_B = 2n\Lambda \quad (2)$$

where λ_B is the Bragg wavelength, n is the effective index of the grating and Λ is the grating period.

Several sensitive layers such as carbon nanotubes (CNT) and graphene oxide [63] have been coated onto FBG to improve the sensor sensitivity. For instance, the CNTs coated produced higher sensitivity as compared to conventional FBG with 31 pm/%RH [64]. The device is depicted in Figure 7 (b).

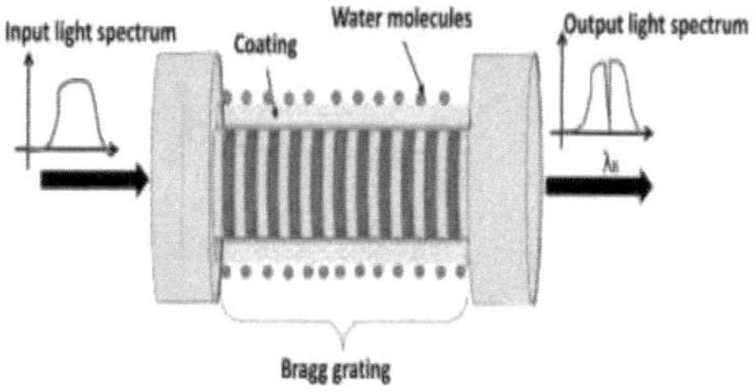

(a) (b)

Figure 7. (a) D-shape coated by tungsten disulfide coating and (b) the SEM image of the coating layer [60], (b) Optical fiber humidity sensor based on fiber Bragg gratings [63]

Modal interferometer can be formed when difference effective refractive indices of different fiber modes produce interferometric phase difference [65]. Numerous topological structures such as Mach-Zehnder interferometers (MZI) could be realized by splicing different types of fibers in a hybrid structure [66]. It requires two different optical paths to generate the interference. One is the core of the optical fiber and the other path is the cladding. The cladding modes would guide propagate signal through the cladding to generate a modal interferometer based on MZI [58]. One of the structure is double in-line adiabatically tapered fibers as shown in Figure 8 [67]. The first tapered region causes fundamental mode to diffract and excited the cladding modes. The effective refractive indices differences between the core and cladding modes produce phase shift. As the structure exposed to higher %RH, the effective RI of the cladding modes increase while the RI of the core mode maintain and lead to a greater phase shift. Performance comparison of some of microfiber based humidity sensors are listed in Table 1.

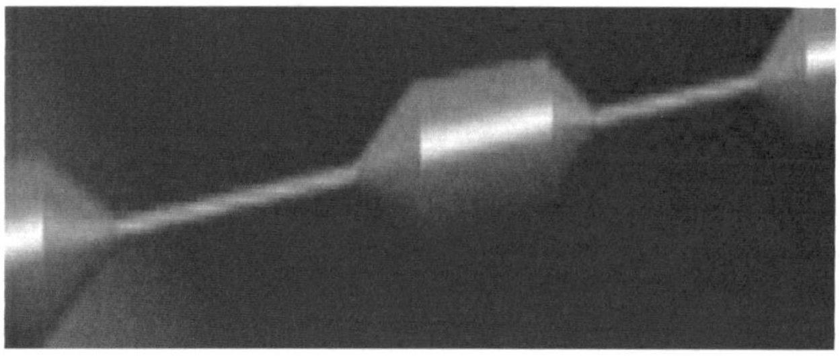

Figure 8. Double in-line adiabatically tapered fibers structures to form Mach-Zehnder interferometer [67]

Table 1. Comparative table of several microfibers based humidity sensor

No	Type	Principle/Method	Performances	References
1	Microfiber bottle resonator	Single & double bottle	Sensitivity of 0.061 dBm/%RH	[54]
2	Strain induced Bragg wavelength	Coated with polyimide	Sensitivity of 2.1 pm/%RH	[68]
3	Microbottle	Integrate with microfiber	Sensitivity of 0.0487 dBm/%RH	[56]
4	PMMA microfiber	Doped with agarose gel	Sensitivity of 0.1 µw/%RH	[69]
5	Microbottle resonator	Coated with PVA	Sensitivity of 0.1311 dBm/%RH	[70]
6	Microfiber with glass substrate	Coated with ZnO nanorods	Sensitivity of 0.0527 dBm/%RH	[71]
7	Hybrid SMF	Coated with PVA	Sensitivity of 1.994 µW/%RH	[72]
8	Strain induced FBG	Coated with polyimide	Sensitivity of 13.6 pm/%RH	[73]
9	SMF microfiber	Laid on glass substrate	Sensitivity of 14.2 pm/%RH	[74]
10	Etched FBG	Coated with CNT	Sensitivity of 31 pm/&RH	[64]
11	Microloop	Coated with ZnO nanorods	Sensitivity of 0.2774 dBm/%RH	[75]
12	Microknot	PMMA microfiber interferometric	Sensitivity of 8.8 pm/%RH	[76]
13	Microtoroid+tapes	Poly (N isopropylacrylamide)	Sensitivity of 13 pm/%RH	[77]
14	Microsphere resonator	Integrated with ZnO coated glass	Sensitivity of 0.0748 dBm/%RH	[78]

6. Conclusion

This paper has summarized most of the related research works pertaining to optical microfiber sensor. It could bring a new glance on the potential research ideas in this field. Since the optical sensor has superiority as compared to electronic sensor such as immunity to the electromagnetic interference, advancement in this field is highly encourage. It expected that this paper could motivate researchers to improvise current optical microfiber sensor towards more sensitive, simple and low cost sensing device.

References

[1] Udd E and Spillman Jr W B 2011 *Fiber optic sensors: an introduction for engineers and scientists*: John Wiley & Sons)

[2] Johari M A M, Jali M H B, Yusof H H B M, Rahim H R B A, Ahmad A B, Khudus M I M A and Harun S W 2021 Polyvinyl alcohol coating microbottle resonator on whispering gallery modes for ethanol liquid sensor *Optics & Laser Technology* **143** 107379

[3] Johari M A M, Sharif M, Jali M H B, Yusof H H B M, Rahim H R B A, Ahmad A B, Khudus M I M A and Harun S W 2021 Effect of polyvinyl alcohol coating microbottle resonator for sodium hypochlorite concentration sensing *Optik* **242** 166824

[4] Jali M H, Rahim H R A, Johari M A M, Ali U, Johari S H, Mohamed H, Harun S W and Yasin M 2021 Formaldehyde sensor with enhanced performance using microsphere resonator-coupled ZnO nanorods coated glass *Optics & Laser Technology* **139** 106853

[5] Kumar R, Leng Y, Liu B, Zhou J, Shao L, Yuan J, Fan X, Wan S, Wu T and Liu J 2019 Ultrasensitive biosensor based on magnetic microspheres enhanced microfiber interferometer *Biosensors and Bioelectronics* **145** 111563
[6] Jali M H, Rahim H R A, Johari M A M, Yusof H H M, Rahman B, Harun S W and Yasin M 2019 Formaldehyde sensing using ZnO nanorods coated glass integrated with microfiber *Optics & Laser Technology* **120** 105750
[7] Lou J, Wang Y and Tong L 2014 Microfiber optical sensors: A review *Sensors* **14** 5823-44
[8] Tong L and Sumetsky M 2011 *Subwavelength and nanometer diameter optical fibers*: Springer Science & Business Media)
[9] Chen G Y, Ding M, Newson T and Brambilla G 2013 A review of microfiber and nanofiber based optical sensors *The Open Optics Journal* **7**
[10] Tang J, Zhou J, Guan J, Long S, Yu J, Guan H, Lu H, Luo Y, Zhang J and Chen Z 2017 Fabrication of side-polished single mode-multimode-single mode fiber and its characteristics of refractive index sensing *IEEE Journal of Selected Topics in Quantum Electronics* **23** 238-45
[11] Bobb L C, Shankar P and Krumboltz H D 1990 Bending effects in biconically tapered single-mode fibers *Journal of lightwave technology* **8** 1084-90
[12] Brambilla G J J o O 2010 Optical fibre nanowires and microwires: a review **12** 043001
[13] Wu X and Tong L 2013 Optical microfibers and nanofibers *Nanophotonics* **2** 407-28
[14] Sumetsky M, Dulashko Y, Fini J M and Hale A 2005 Optical microfiber loop resonator *Applied Physics Letters* **86** 161108
[15] Zhai G and Tong L J O e 2007 Roughness-induced radiation losses in optical micro or nanofibers **15** 13805-16
[16] Sumetsky M J O l 2006 How thin can a microfiber be and still guide light? **31** 870-2
[17] Brambilla G, Xu F and Feng X J E L 2006 Fabrication of optical fibre nanowires and their optical and mechanical characterisation **42** 517-9
[18] Le Kien F, Balykin V and Hakuta K 2006 Angular momentum of light in an optical nanofiber *Physical Review A* **73** 053823
[19] Tong L, Lou J and Mazur E 2004 Single-mode guiding properties of subwavelength-diameter silica and silicon wire waveguides *Optics Express* **12** 1025-35
[20] Brambilla G 2010 Optical fibre nanowires and microwires: a review *Journal of Optics* **12** 043001
[21] Zibaii M, Latifi H, Karami M, Gholami M, Hosseini S and Ghezelayagh M 2010 Non-adiabatic tapered optical fiber sensor for measuring the interaction between α-amino acids in aqueous carbohydrate solution *Measurement Science and Technology* **21** 105801
[22] Ravets S, Hoffman J, Orozco L, Rolston S, Beadie G and Fatemi F 2013 A low-loss photonic silica nanofiber for higher-order modes *Optics express* **21** 18325-35
[23] Tong L, Lou J, Ye Z, Svacha G T and Mazur E J N 2005 Self-modulated taper drawing of silica nanowires **16** 1445
[24] Bilodeau F, Hill K, Faucher S and Johnson D J J o L T 1988 Low-loss highly overcoupled fused couplers: Fabrication and sensitivity to external pressure **6** 1476-82
[25] Birks T A and Li Y W J J o L T 1992 The shape of fiber tapers **10** 432-8
[26] Brambilla G, Koizumi F, Feng X and Richardson D J E L 2005 Compound-glass optical nanowires **41** 400-2
[27] Sumetsky M, Dulashko Y, Fini J M and Hale A J A P L 2005 Optical microfiber loop resonator **86** 161108
[28] Tong L, Hu L, Zhang J, Qiu J, Yang Q, Lou J, Shen Y, He J and Ye Z J O E 2006 Photonic nanowires directly drawn from bulk glasses **14** 82-7
[29] Kieu K Q and Mansuripur M 2006 Biconical fiber taper sensors *IEEE Photonics Technology Letters* **18** 2239-41
[30] Corres J M, Arregui F J and Matias I R 2006 Design of humidity sensors based on tapered optical fibers *Journal of Lightwave Technology* **24** 4329-36
[31] Kakarantzas G, Dimmick T, Birks T, Le Roux R and Russell P S J 2001 Miniature all-fiber devices

based on CO 2 laser microstructuring of tapered fibers *Optics Letters* **26** 1137-9

[32] Boucouvalas A and Georgiou G 1985 Biconical taper coaxial coupler filter *Electronics Letters* **21** 1033-4

[33] Dimmick T E, Kakarantzas G, Birks T A and Russell P S J 1999 Carbon dioxide laser fabrication of fused-fiber couplers and tapers *Applied Optics* **38** 6845-8

[34] Bohren C F and Huffman D R 2008 *Absorption and scattering of light by small particles*: John Wiley & Sons)

[35] Coillet A, Cluzel B, Vienne G, Grelu P and de Fornel F 2010 Near-field characterization of glass microfibers on a low-index substrate *Applied Physics B* **101** 291-5

[36] Elosua C, Arregui F J, Villar I D, Ruiz-Zamarreño C, Corres J M, Bariain C, Goicoechea J, Hernaez M, Rivero P J and Socorro A B 2017 Micro and nanostructured materials for the development of optical fibre sensors *Sensors* **17** 2312

[37] Brambilla G 2010 Optical fibre nanotaper sensors *Optical Fiber Technology* **16** 331-42

[38] Jali M H, Rahim H R A, Hamid S S, Johari M A M, Yusof H H M, Thokchom S, Harun S W, Khasanah M and Yasin M 2019 Microfiber loop resonator for formaldehyde liquid sensing *Optik* **196** 163174

[39] Vienne G, Grelu P, Pan X, Li Y and Tong L 2008 Theoretical study of microfiber resonator devices exploiting a phase shift *Journal of optics A: pure and applied optics* **10** 025303

[40] Guo X and Tong L 2008 Supported microfiber loops for optical sensing *Optics Express* **16** 14429-34

[41] Xu F and Brambilla G J A P L 2008 Demonstration of a refractometric sensor based on optical microfiber coil resonator **92** 101126

[42] Brambilla G J O F T 2010 Optical fibre nanotaper sensors **16** 331-42

[43] Pöllinger M, O'Shea D, Warken F and Rauschenbeutel A J P r l 2009 Ultrahigh-Q tunable whispering-gallery-mode microresonator **103** 053901

[44] Knight J C, Cheung G, Jacques F and Birks T J O l 1997 Phase-matched excitation of whispering-gallery-mode resonances by a fiber taper **22** 1129-31

[45] Fan X, White I M, Shopova S I, Zhu H, Suter J D and Sun Y J a c a 2008 Sensitive optical biosensors for unlabeled targets: A review **620** 8-26

[46] Vollmer F, Braun D, Libchaber A, Khoshsima M, Teraoka I and Arnold S J A p l 2002 Protein detection by optical shift of a resonant microcavity **80** 4057-9

[47] Noto M, Vollmer F, Keng D, Teraoka I and Arnold S J O l 2005 Nanolayer characterization through wavelength multiplexing of a microsphere resonator **30** 510-2

[48] Vollmer F, Arnold S and Keng D J P o t N A o S 2008 Single virus detection from the reactive shift of a whispering-gallery mode **105** 20701-4

[49] Alwis L, Sun T and Grattan K 2013 Optical fibre-based sensor technology for humidity and moisture measurement: Review of recent progress *Measurement* **46** 4052-74

[50] Yuan W, Wang X, Nie Y, Li J and Sang M 2015 Optical relative humidity sensor with high sensitivity based on a polyimide-coated symmetrical metal-cladding waveguide *Applied optics* **54** 834-8

[51] Hosseini Z and Mortezaali A 2015 Room temperature H2S gas sensor based on rather aligned ZnO nanorods with flower-like structures *Sensors and Actuators B: Chemical* **207** 865-71

[52] Mani G K and Rayappan J B B 2015 A highly selective and wide range ammonia sensor—Nanostructured ZnO: Co thin film *Materials Science and Engineering: B* **191** 41-50

[53] Kannan P K, Saraswathi R and Rayappan J B B 2010 A highly sensitive humidity sensor based on DC reactive magnetron sputtered zinc oxide thin film *Sensors and Actuators A: Physical* **164** 8-14

[54] Johari M A M, Khudus M I M A, Jali M H B, Al Noman A and Harun S W 2018 Effect of size on single and double optical microbottle resonator humidity sensors *Sensors and Actuators A: Physical* **284** 286-91

[55] Wang Y, Yeow J T and Chen L-Y 2009 *Recent advances in sensing technology*: Springer) pp 257-

69
[56] Johari M A M, Khudus M I M A, Jali M H B, Al Noman A and Harun S W 2019 Whispering gallery modes on optical micro-bottle resonator for humidity sensor application *Optik* **185** 558-65
[57] Lokman A, Nodehi S, Batumalay M, Arof H, Ahmad H and Harun S W 2014 Optical fiber humidity sensor based on a tapered fiber with hydroxyethylcellulose/polyvinylidenefluoride composite *Microwave and Optical Technology Letters* **56** 380-2
[58] Ascorbe J, Corres J M, Arregui F J and Matias I R 2017 Recent developments in fiber optics humidity sensors *Sensors* **17** 893
[59] Yusof H H M, Jali M H, Johari M A M, Dimyati K, Harun S W, Khasanah M and Yasin M 2019 Detection of formaldehyde vapor using glass substrate coated with zinc oxide nanorods *IEEE Photonics Journal* **11** 1-9
[60] Luo Y, Chen C, Xia K, Peng S, Guan H, Tang J, Lu H, Yu J, Zhang J and Xiao Y 2016 Tungsten disulfide (WS 2) based all-fiber-optic humidity sensor *Optics express* **24** 8956-66
[61] Xiao Y, Zhang J, Cai X, Tan S, Yu J, Lu H, Luo Y, Liao G, Li S and Tang J 2014 Reduced graphene oxide for fiber-optic humidity sensing *Optics express* **22** 31555-67
[62] Zhou Z, Liu W, Huang Y, Wang H, Jianping H, Huang M and Jinping O 2012 Optical fiber Bragg grating sensor assembly for 3D strain monitoring and its case study in highway pavement *Mechanical Systems and Signal Processing* **28** 36-49
[63] Wang Y, Shen C, Lou W, Shentu F, Zhong C, Dong X and Tong L 2016 Fiber optic relative humidity sensor based on the tilted fiber Bragg grating coated with graphene oxide *Applied Physics Letters* **109** 031107
[64] Shivananju B N, Yamdagni S, Fazuldeen R, Kumar A K S, Nithin S P, Varma M M and Asokan S 2014 Highly sensitive carbon nanotubes coated etched fiber Bragg grating sensor for humidity sensing *IEEE Sensors Journal* **14** 2615-9
[65] Wu Q, Semenova Y, Mathew J, Wang P and Farrell G 2011 Humidity sensor based on a single-mode hetero-core fiber structure *Optics Letters* **36** 1752-4
[66] Wang Y, Shen C, Lou W and Shentu F 2016 Fiber optic humidity sensor based on the graphene oxide/PVA composite film *Optics Communications* **372** 229-34
[67] Soltanian M R K, Sharbirin A S, Ariannejad M, Amiri I, De La Rue R, Brambilla G, Rahman B A, Grattan K and Ahmad H 2016 Variable waist-diameter Mach–Zehnder tapered-fiber interferometer as humidity and temperature sensor *IEEE sensors Journal* **16** 5987-92
[68] Berruti G, Consales M, Giordano M, Sansone L, Petagna P, Buontempo S, Breglio G and Cusano A 2013 Radiation hard humidity sensors for high energy physics applications using polyimide-coated fiber Bragg gratings sensors *Sensors and Actuators B: Chemical* **177** 94-102
[69] Irawati N, Rahman H A, Yasin M, Al-Askari S, Hamida B A, Ahmad H and Harun S W 2017 Relative humidity sensing using a PMMA doped agarose gel microfiber *Journal of Lightwave Technology* **35** 3940-4
[70] Johari M A M, Pour M P, Al Noman A, Khudus M I M A, Jali M H B, Maslinda M, Ali U U M and Harun S W 2020 Effect of PMMA and PVA coating on the performance of optical microbottle resonator humidity sensors *Microwave and Optical Technology Letters* **62** 993-8
[71] Jali M H, Rahim H R A, Johari M A M, Yusof H H M, Ahmad A, Thokchom S, Dimyati K and Harun S W 2021 Humidity sensing using microfiber-ZnO nanorods coated glass structure *Optik* **238** 166715
[72] Li T, Dong X, Chan C C, Zhao C-L and Zu P 2011 Humidity sensor based on a multimode-fiber taper coated with polyvinyl alcohol interacting with a fiber Bragg grating *IEEE Sensors Journal* **12** 2205-8
[73] Lin Y, Gong Y, Wu Y and Wu H 2015 Polyimide-coated fiber Bragg grating for relative humidity sensing *Photonic sensors* **5** 60-6
[74] Jali M H, Rahim H R A, Johari M A M, Sharif M M, Johari S H, Thokchom S, Mohamed H, Yasin M and Harun S W 2021 Applied whispering gallery modes on ZnO nanorods coated glass for

humidity sensing application *Optoelectronics Letters* **17** 298-301

[75] Zain H A, Jali M H, Rahim H R A, Johari M A M, Yusof H H M, Thokchom S, Yasin M and Harun S W 2020 ZnO nanorods coated microfiber loop resonator for relative humidity sensing *Optical Fiber Technology* **54** 102080

[76] Wu Y, Zhang T, Rao Y and Gong Y 2011 Miniature interferometric humidity sensors based on silica/polymer microfiber knot resonators *Sensors and Actuators B: Chemical* **155** 258-63

[77] Mehrabani S, Kwong P, Gupta M and Armani A M 2013 Hybrid microcavity humidity sensor *Applied Physics Letters* **102** 241101

[78] Jali M H, Rahim H R A, Johari M A M, Ahmad A, Yusof H H M, Johari S H, Baharom M F, Thokchom S, Dimyati K and Harun S W 2021 Integrating microsphere resonator and ZnO nanorods coated glass for humidity sensing application *Optics & Laser Technology* **143** 107356

Acknowledgement
The authors would like to thank Universiti Teknikal Malaysia Melaka (UTeM) for the financial support.

AUTHOR INDEX